Drones and Responsibility

How does the use of military drones affect the legal, political, and moral responsibility of different actors involved in their deployment and design? This volume offers a fresh contribution to the ethics of drone warfare by providing, for the first time, a systematic interdisciplinary discussion of different responsibility issues raised by military drones. The book discusses four main sets of questions: First, from a legal point of view, we analyse the ways in which the use of drones makes the attribution of criminal responsibility to individuals for war crimes more complicated and what adjustments may be required in international criminal law and in military practices to avoid 'responsibility gaps' in warfare. From a moral and political perspective, the volume looks at the conditions under which the use of military drones by states is impermissible, permissible, or even obligatory and what the responsibilities of a state in the use of drones towards both its citizens and potential targets are. From a socio-technical perspective, what kind of new human machine interaction might (and should) drones bring and which new kinds of shared agency and responsibility? Finally, we ask how the use of drones changes our conception of agency and responsibility. The book will be of interest to scholars and students in (military) ethics and to those in law, politics and the military involved in the design, deployment and evaluation of military drones.

Ezio Di Nucci is Associate Professor of Medical Ethics at the University of Copenhagen, having previously taught at the University of Edinburgh (where he received his PhD in 2008), the University of Stirling, University College Dublin and the University of Duisburg-Essen (where he received his *Habilitation* in 2014). Ezio works mainly in ethics, bioethics and the philosophy of action.

Filippo Santoni de Sio is Assistant Professor of Philosophy and Ethics of Technology at Delft University of Technology. He received his PhD in Philosophy at the University of Turin in 2008. He has already published one monograph, one edited collection and more than thirty papers on moral and legal responsibility, the ethics of cognitive enhancement, and robot ethics. He is in the organizing committee of the Foundation for Responsible Robotics.

Emerging Technologies, Ethics, and International Affairs

Series editors:

Steven J. Barela, *University of Geneva, Switzerland*
Jai C. Galliott, *The University of New South Wales, Australia*
Avery Plaw, *University of Massachusetts, USA*
Katina Michael, *University of Wollongong, Australia*

This series examines the crucial ethical, legal, and public policy questions arising from or exacerbated by the design, development, and eventual adoption of new technologies across all related fields, from education and engineering to medicine and military affairs.

The books revolve around two key themes:

- Moral issues in research, engineering and design.
- Ethical, legal and political/policy issues in the use and regulation of technology.

This series encourages submission of cutting-edge research monographs and edited collections with a particular focus on forward-looking ideas concerning innovative or as yet undeveloped technologies. Whilst there is an expectation that authors will be well grounded in philosophy, law or political science, consideration will be given to future-oriented works that cross these disciplinary boundaries. The interdisciplinary nature of the series editorial team offers the best possible examination of works that address the ethical, legal, and social implications of emerging technologies.

Forthcoming titles:

The Rise of the Good Drone
Edited by Kristin Bergtora Sandvik and Maria Gabrielsen Jumbert

Military Neuroscience and the Coming Age of Neurowarfare
Armin Krishnan

Emerging Technologies in Forensic Science
Edited by Ronn Johnson

Drones and Responsibility

Legal, Philosophical, and Sociotechnical Perspectives on Remotely Controlled Weapons

Edited by
Ezio Di Nucci and
Filippo Santoni de Sio

Routledge
Taylor & Francis Group
LONDON AND NEW YORK

First published 2016 by Routledge

2 Park Square, Milton Park, Abingdon, Oxfordshire OX14 4RN
52 Vanderbilt Avenue, New York, NY 10017

Routledge is an imprint of the Taylor & Francis Group, an informa business

First issued in paperback 2019

British Library Cataloguing in Publication Data
A catalogue record for this book is available from the British Library

Library of Congress Cataloging-in-Publication Data
Names: Nucci, Ezio Di, editor. | Sio, Filippo Santoni de, editor.
Title: Drones and responsibility : legal, philosophical and socio-technical perspectives on remotely controlled weapons / by Ezio Di Nucci and Filippo Santoni de Sio.
Description: Farnham, Surrey, UK England ; Burlington, VT : Ashgate, 2016. | Series: Emerging technologies, ethics and international affairs | Includes bibliographical references and index.
Identifiers: LCCN 2015039243 | ISBN 9781472456724 (hardback : alk. paper)
Subjects: LCSH: Uninhabited combat aerial vehicles (International law) | Uninhabited combat aerial vehicles—Moral and ethical aspects.
Classification: LCC KZ6687 .D765 2016 | DDC 341.6/3—dc23

ISBN: 978-1-4724-5672-4 (hbk)
ISBN: 978-1-138-39066-9 (pbk)

Typeset in Times New Roman
by Apex CoVantage, LLC

Contents

Contributors

Ezio Di Nucci is Associate Professor of Medical Ethics at the University of Copenhagen.

Michael Funk is a researcher at the Department of Philosophy of the University of Vienna.

Tjerk de Greef is a Visiting Researcher at Delft University of Technology.

Bernhard Irrgang is Professor and holds the Chair at the Department for Philosophy of Technology at TU Dresden.

Asa Kasher is Laura Schwarz-Kipp Professor Emeritus of Professional Ethics and Philosophy of Practice and Emeritus Professor of Philosophy, Tel Aviv University, and Professor of Philosophy, Shalem Academic Center, Jerusalem.

Jesse Kirkpatrick is Assistant Director, Institute for Philosophy and Public Policy at George Mason University.

Bernhard Koch is Deputy Director of the Institute for Theology and Peace in Hamburg and Visiting Lecturer of Philosophy at the Goethe University Frankfurt/Main.

Silvio Leuteritz is a PhD student at the Department for Philosophy of Technology at TU Dresden.

Alex Leveringhaus is a Postdoctoral Research Fellow at the University of Oxford.

Chantal Meloni is Senior Researcher in Criminal Law and Adjunct Professor of International Criminal Law at the University of Milan and Guest Researcher at the Humboldt University of Berlin.

Nikil Mukerji is a Postdoctoral Researcher in Philosophy at Ludwig-Maximilians-Universität München.

Vincent C. Müller is Professor of Philosophy at Anatolia College.

Michael Nagenborg is Assistant Professor in Philosophy of Technology at the University of Twente.

Filippo Santoni de Sio is Assistant Professor of Philosophy and Ethics of Technology at Delft University of Technology.

Dan Saxon is Assistant Professor in Law at Leiden University.

Aimee van Wynsberghe is Assistant Professor in Ethics of Technology at the University of Twente.

Preface

Drones are rapidly changing the way in which wars are fought. It is unsurprising, then, that the ethical debate on military drones has been intensifying over the last few years. This book aims to contribute to that debate by looking at one particular ethical issue: responsibility. Our volume investigates different forms of responsibility (moral responsibility, legal responsibility, and political responsibility) from a broad interdisciplinary perspective, bringing together philosophers, legal scholars, engineers, and social scientists. The volume has its origins in a workshop that the editors organized in the spring of 2014 in Delft: we are very grateful to all the participants to that workshop and to Delft University of Technology for making the meeting possible. We are proud to have managed to bring together at that workshop very different scholars such as, for example, Dan Saxon, former prosecutor at the International Criminal Tribunal for the former Yugoslavia in The Hague, and Asa Kasher, a philosopher famous for having written the ethics code for the Israel Defense Forces. We think having offered a joint platform to such different points of view is a big achievement of that meeting and of this volume. Also, we have strived to provide each author with as much access as possible to what the others were writing, so that the reader will find that, often, the different chapters "talk" to each other. Finally, a note on the terminology: we have decided to use the term "drone" among the many alternatives in circulation (such as UAV [unmanned aerial vehicle] or RPA [remotely piloted aircraft] for example) for simplicity's sake: it is the most used in the public debate; however, much of what is said about drones in this book may also apply to other, non-flying, remotely controlled or semiautomated weapon systems. Also, if not otherwise specified, the term "drone" throughout the book refers to military drones.

Ezio Di Nucci and Filippo Santoni de Sio have equally contributed to the preparation of this book.

Copenhagen and Delft
10 March 2016

Drones and Responsibility

Mapping the Field

Filippo Santoni de Sio and Ezio Di Nucci

1. Introduction

In the past few years, not only political campaigners and media but also academic scholars have engaged in a discussion on the ethics of drone wars. The academic ethical debate on military drones has mainly focused on two topics. On the one hand, the general ethical question has been posed as to whether it is alright to use remotely controlled weapons in (lethal) military operations (Strawser 2013). On the other hand, one more specific issue has been discussed: whether the use of machines that enjoy some level of autonomy in military operations may lead to ethically unacceptable "responsibility gaps" – circumstances in which an untoward event occurs and no human agent can legitimately be held responsible for it (Matthias 2004; Sparrow 2007). The present volume aims to analyze in more depth the relationship between the use of drones in military operations and various issues of responsibility. It has something in common with both of the existing debates, but it also significantly exceeds them in scope. The volume is related to the general debate on the permissibility of the use of military drones insofar as one of the issues of responsibility it discusses is whether and under which conditions (democratic) states have a responsibility (a duty) to use military drones, for instance, in order to put fewer soldiers in harm's way; this volume is clearly related also to the responsibility gap debate, insofar as this debate concerns *one* possible issue of responsibility raised by the use of military drones. However, the scope of this book is broader. We think that the issue of responsibility gaps as traditionally conceived of in the philosophical debate is overrated and that there are a lot more interesting things to say about responsibility in relation to drones than those present in that philosophical debate. As it were, even though some of the chapters do address this topic, moving beyond the responsibility gap debate is one of the main goals of this volume. In the following section, we explain why we don't believe in the responsibility gap; in the rest of this chapter, we then go on to describe the different issues of responsibility that we deem relevant and that are addressed in this volume.

2. State of the Art: Responsibility and the Responsibility Gap

Talk of the responsibility gap (Matthias 2004; Sparrow 2007) refers to the idea that, when technology that has a certain degree of autonomy is deployed, there is a risk that no human will be responsible for what this autonomous or semiautonomous technology will do, for example, in case such technology should happen to malfunction and create negative unwanted effects. This supposed responsibility gap is then taken to be an ethical reason against deploying autonomous or semi-autonomous technology.

Sparrow's argument is that there is no plausible candidate for the bearer of responsibility (say, for war crimes): the programmer is not a plausible candidate; the commanding officer is not a plausible candidate; and the machine itself is also not a plausible candidate: "the impossibility of punishing the machine means that we cannot hold the machine responsible. We can insist that the officer who orders their use be held responsible for their actions, but only at the cost of allowing that they should sometimes be held entirely responsible for actions over which they had no control" (Sparrow 2007: 74).

We agree with Sparrow that the machine itself cannot be the bearer of responsibility. We also agree with Sparrow that there will be at least some cases where the programmers cannot be made legitimately responsible for what has gone wrong, either because they had actually mentioned the risk to the decision-makers or because the failing could not reasonably have been predicted by programmers.

What is left is the decision-maker (Sparrow's commanding officer). And indeed here is where Sparrow's argument is at its weakest. Sparrow argues that the commanding officer cannot be held responsible for everything the machines will do because "the autonomy of the machine implies that its orders do not determine (although they obviously influence) its actions" (Sparrow 2007: 71).

First of all, this principle – understood as a necessary condition for responsibility according to which a commanding officer or decision-maker can be responsible only if his orders determine the relevant actions that will be carried out – is implausible. Firstly, because whether we live in a deterministic or indeterministic world, orders alone will never determine actions, whether of humans or robots. So the kind of determination meant cannot be that of the thesis of causal determinism; but even taking the principle to rather mean something about orders being a proximate cause in a reliable chain leading to action, the principle would remain implausible because it would negate responsibility in most cases of ignorance, negligence, and bad luck. Indeed, Sparrow has in mind a particular sense of "moral responsibility" as relying on a strong control condition which, as philosophically defensible as it might be, is certainly not the (only) one used in common morality, let alone in the law. So Sparrow's principle is, as a general necessary condition for responsibility, way too weak.

Moreover, as far as military robots are concerned, that the commanding officer will not be responsible for everything the machine does is not a problem because one may clearly also accept a pluralist view according to which sometimes the programmer is responsible. However, Sparrow's claim is that sometimes the commanding officer is not responsible and no one else is, and that is the problem.

But even if one accepts that the machine's orders do not fully determine its actions, that in turn does not imply that sometimes the commanding officer will be – for exactly that reason – not responsible. For example, it may be that the simple fact that the commanding officer is aware of this problem will be enough for an attribution of responsibility. And further it is plausible to suppose that the decision-makers are obliged to inform themselves on exactly these kinds of risks. Here the context of deployment may make a difference. In the case of the military chain of command, we may be unwilling to make the commanding officer responsible for the malfunctioning of a machine despite her awareness because the commanding officer is herself subject to orders – still, someone else within that chain of command will then be bearer of responsibility.

Decision-makers must inform themselves about the levels of flexibility and responsiveness to the environment of the machines and about possible malfunctioning related to this. In making their deployment decisions at the different levels, decision-makers are or should be aware of this and are therefore responsible for the consequences of their deployment decisions – at least those that could have been reasonably foreseen, where obviously this condition is more challenging when it comes to autonomous robots than with simple instruments or machines.

So there is no responsibility gap insofar as the decision-makers can be held responsible for all the malfunctioning and unpredicted functioning that could have been reasonably predicted. Admittedly, what would have been reasonable to predict is, in this case, possibly more challenging than in more traditional decision making, and future courts and legislators may have a very hard time in making a solid case for malfunctions. Some of these further complications are extensively discussed in the legal part of this volume.

However, some of this complication is not new and should therefore not be overstated. Take, for example, decision-making chains. Even if robots or machines are not involved, attributing responsibility within a complicated decision-making chain such as a state or company is already a very difficult task (van de Poel et al 2015), and often when it comes to legal responsibility, courts do fail due to the sophistication of the decision-making chain. Robots may make this even more difficult, but, again, these difficulties are not new and therefore should not be overstated in the case of robots.

In addition, even when some difficulties are likely to arise, these responsibility gaps are not inevitable. By becoming aware of the possible difficulties of future legislators and courts in attributions of moral and legal responsibility for the malfunctioning of robots, we may and should find ways to grant that future legislators and courts will have resources that current legislators and courts do not possess: the level of sophistication grows on both sides.

Finally, not only legislators but also designers and programmers may contribute to reduce future responsibility gaps: we believe that ethical considerations and values should be taken into account not only at the level of use but also at the early stage of technological design. From this perspective, known in the ethics of technology debate as value-sensitive design, human control and responsibility may be seen as nonfunctional requirements that can and should be met in the design stage of sociotechnical systems.

3. Beyond the Responsibility Gap: A Pluralist and Interdisciplinary Perspective on Drones and Responsibility

One simple reaction to the debate on military drones and responsibility would consist of taking a reductionist perspective: insisting that drones are just (very complicated) tools, so that their use does not detract in any way from the responsibility of the human beings who deploy them. In the same way in which, for instance, the deployment of a gun as opposed to a knife should not make any difference in the attribution of responsibility for a murder to a gunman as opposed to a stabber; similarly, the use of an armed drone rather than other military weapons should not make any difference in the attribution of responsibility for military operations. In the end, so the argument goes, the burden of responsibility for the use of drones in military operations – and the consequence of this – remain on politicians and commanders who decide whether and when to deploy drones and who are thus responsible for the (mis)use or malfunctioning of drones in military operations, as well as the consequences of this (mis)use or malfunction. In this reductionist perspective, the only relevant question about drones and responsibility is therefore whether and under which conditions the use by states of (which) military drones is permissible, impermissible, or obligatory.

We admit that drones are in an important sense just tools, machines to which no moral or political responsibility can be assigned, and we also recognize that the permissibility/impermissibility/obligatoriness of the use of drones by (democratic) states is certainly one very important issue to be discussed. However, we also think that the deployment of drones in military operations *does* pose other interesting questions of responsibility. Admittedly, it may certainly be the case that some or even many of these other questions are not completely new; that is, they are rather variations or complications of more common issues of responsibility already present in the moral, political, legal, or socio-technological reflection on war, technology, and responsibility. In other words, it may be the case that drones introduce only a quantitative not a qualitative change in the continuum of the evolution of military technology.

However, we think there are still good reasons to explore these drones and responsibility issues in some depth. Firstly, some quantitative changes are more relevant than others: for instance, the passage from a knife to a gun can be of more momentum than the passage from the club to the knife. Due to the specific features of different tools, designers, producers, sellers, legislators may have responsibilities for the wrong use of guns that they wouldn't have for the wrong use of knives. Secondly, even if the analysis had to reveal that drones do not pose any significantly new question of responsibility, we think that drones may represent a new interesting case study whose analysis may contribute to a rethinking, a deepening, and possibly an extension of our understanding of responsibility, technology, and warfare – or maybe just of responsibility *tout court*.

In order to map the different issues at stake, we suggest a *pluralist* and *interdisciplinary* approach to drones and responsibility. Our approach is pluralist because, in contrast to the responsibility gap theorists, our focus is not just on individual backward-looking moral blameworthiness for conscious and/or intended and/or rationally deliberated acts. We think that there are other kinds of responsibility to be attributed to different individual and collective actors in different circumstances: for instance, the responsibility of designers for the unintended (wrong) consequences of their technological creations; the responsibility of collective agents like democratic states to give the public an account of their military operations and the outcomes of these; the moral and legal responsibilities deriving from the occupation of a particular position, like that of a military commander for the misbehaviour of his soldiers.

We take an interdisciplinary approach, as we think that, in order to map as many interesting issues of responsibility as possible, contributions from different disciplinary perspectives are required – not only philosophy but also law and science/engineering.

4. Varieties of Responsibility

In order to give a comprehensive map of the different problems of responsibility raised by the use of military drones, we will firstly present the different kinds of responsibility as well as the different actors potentially involved in responsibility attributions for the use of drones.

As for responsibility, a first important distinction is that between moral, legal, and political responsibility. By *moral* responsibility, we mean the responsibilities that arise by simply being humans acting in a space with other humans. A paradigmatic example of moral responsibility is the blameworthiness for stealing something or intentionally injuring someone. By *legal* and *political* responsibility, we mean those specific responsibilities that come into existence due to the presence of a particular legal or political system of rules. A paradigmatic example of legal responsibility would be A's liability to pay a certain sum of money to B, a liability grounded in a valid legal contract existing between A and B. A paradigmatic example of political responsibility would be the duty of a democratic government to give their citizen an account of their activities. While we think that legal and political responsibility are often also a kind of moral responsibility, we think it is important to keep in mind the difference between them and simpler cases of moral responsibility.

In addition, moral, political, and legal responsibility can be backward-looking or forward-looking. *Backward-looking* responsibility designates the normative position of an actor in relation to facts that have already occurred, for instance someone being subject to moral blame (or praise) for one particular past action of them; *forward-looking* responsibility concerns the normative position of an actor in relation to future actions and states of affair, for instance, a government being subject to an obligation to take preventive steps in order to avoid the occurrence of certain accidents on a given territory.

Backward-looking responsibility (be it moral, political, or legal) comes in at least five different kinds: capacity-responsibility, accountability, liability, causal responsibility, and role-responsibility (see Hart 1968; Vincent 2011).

According to *capacity-responsibility*, one is responsible if they are the kind of agents who possess the prerequisites to be attributed (a certain kind of) backward-looking responsibility. A typical example of capacity-responsibility is a person's ability to have meaningful moral interactions with others, that is, being a sane adult able to understand the nature and meanings of his actions and their effects on others.

By *accountability*, we mean the normative position of who ought to report and explain – indeed, to give an account – of something that has happened. Typical examples of accountability are an individual person having the obligation to give reasons for a certain morally relevant action she made – in order to be attributed praise or blame or simply to help others make sense of what happened and a democratic government being obliged to inform their citizens about its activities and the results of them.

Liability designates the position of who is subject to negative (moral, legal, political) consequences for their actions and omissions. Typical examples would be moral blame and legal punishment for individuals' wrong actions. Liability has in turn at least two main forms: liability for intended actions and liability for unintended but still culpable actions (negligent actions). It is important to keep in mind that not all wrong actions attract liability. Standard ways to avoid liability for wrong behaviour are exemptions, justifications, and excuses. *Exemptions* make subjects in general not suitable to take part in the attribution of liability game; one typical exemption from individual liability is the lack of the relevant capacity (for instance, an individual being affected by a serious mental disorder affecting her ability to perceive the world correctly). *Justifications* work by making a *prima facie* wrong behavior, all things considered, permissible by pointing to the presence of some exceptional circumstances; a typical example is self-defence as a justification for a killing. *Excuses* work by making a *prima facie* liable actor, all things considered, non-blameworthy by pointing to some particular circumstances of the action; typical examples are non-culpable ignorance of relevant circumstances and being coerced. In a nutshell, exemption depends on the general status of the agent, justification depends on the all-things-considered permissibility of the behaviour, excuses on the all-things-considered blameworthiness of the actor.

Causal responsibility designates the position of someone who has acted in such a way as to make – possibly without intention or even without knowledge – a substantial causal contribution to a certain (negative) outcome. A typical example would be that of a scientist unknowingly contributing to the research that eventually leads to the production and use of a weapon of mass destruction.

Finally, *role-responsibility* designates the position of someone who is liable for untoward events because of occupying a certain position, thus independently of her having directly caused them with her behaviour. A typical example of role-responsibility would be the liability of parents for their children's behaviour or that of a commander for the misdeeds of the soldiers under his command.

As for the different actors potentially involved in various issues of responsibility as raised by the use of military drones, we think that at least the following have to be considered: democratic states, individual military personnel (commanders and drone operators), designers, and programmers.

5. Drones and Responsibility: Legal, Philosophical, and Sociotechnical Perspectives

By considering the different actors involved, we may identify at least three set of issues of responsibility in relation to the use of drones: the responsibilities of democratic states, the responsibilities of individual actors using military drones (that is, military personnel), and the responsibilities of designers and programmers of drones.

By using the conceptual distinctions about responsibility and the list of different actors of the previous section, we can provide the following provisional list of different responsibility questions potentially raised by the use of military drones:

- (When) are states *justified* and thus non-liable in using drones for extraterritorial killings? When (if ever) do these killings count as self-defence?
- May the use of drones for extraterritorial killings unduly reduce states' *accountability* for these killings?
- May the use of drones unduly reduce the states' ability to discharge the duties deriving from their *causal responsibility*, such as the duty to compensate or apologize to the innocent victims of lawful drone strikes?
- By keeping soldiers away from the battlefield, are states preventing soldiers from discharging their *responsibility as combatants*?
- Do states have a *responsibility to recognize drone operators as combatants* and to provide them with adequate medical and psychological help?
- May the use of drones make it more difficult to hold *morally* and *legally liable* individuals who directly support with their acts the commission of violations of the laws of war, for instance, by providing soldiers with a new kind of excuse – reliance on the machine?
- May the use of drones make it more difficult to legitimately attribute *role-responsibilities*, such as the "superior" or "command" legal responsibility for war crimes?
- Is it *in principle* morally wrong to delegate (part of) the decision about killing to artificial agents that are not responsible because they lack *the capacity to respond to moral reasons* in the same way as humans do? Or is delegation of these activities OK insofar as machines are as reliable as humans in performing certain tasks?
- Which moral responsibilities derive from *human capacity* for morally autonomous action? How should these human responsibilities be preserved by *design* in the use of military drones?
- How can drones be *designed as to be "accountable"* (i.e. their behaviour being transparent) to human operators?
- What are the *responsibilities of drone designers towards society*? How should designers and programmers make sure that the transition to the civilian uses of drones does not bring unwanted ethical consequences?

5.1 States' Responsibilities

In order to defend the ethical acceptability of the 2014 Israeli military operation in the Gaza Strip, the philosopher Asa Kasher has argued that democratic states have at least two kinds of forward-looking responsibility: they have a political duty to provide its citizens with protection of their life, well-being, and liberty (the Duty of Self Defence), but they also have a moral duty to respect human dignity in every activity (the principle of Human Dignity Protection). Kasher claims that the principle of Human Dignity Protection poses a burden on democratic states to observe strict regulations and clear limitations in the use of armed drones for targeted killings outside their territory. In particular, the use of lethal force by the state must be limited to cases where there is clear evidence that the targeted person represents an imminent threat to the life of the state's citizens; in addition, these operations must be regulated by the traditional war ethics principles of proportionality and necessity. However, according to Kasher, if all these conditions are met and all the constraints observed, the Duty of Self Defence makes the use of military force by the state for targeted killings not only morally permissible but also morally obligatory. Kasher relies on the traditional justification of self-defence to claim that some killings are permissible and not morally blameworthy and that their perpetrators are not legally liable for them (Kasher 2014).[1]

However, some objections to Kasher's view can be raised from an ethical as well as from a legal perspective. Firstly, as remarked in this volume by Bernard Koch, Kasher's interpretation of the principle of self-defence can be challenged from a moral point of view. On the one hand, the principle of self-defence is commonly thought to justify a killing only if at the moment of the killing the victim was posing an immediate lethal threat to the killer; yet it seems that many if not all of those whom Kasher would consider the legitimate target of a state's lethal drone attack cannot be seen as an immediate threat to the life of the state's citizens in the paradigmatic sense typical of the doctrine of self-defence. Whilst in the standard cases of justified killings in self-defence, the victim is killed while she is unlawfully attacking the killer, targets of drone attacks are normally individuals living outside the state, who at the time of the drone attack are thought to be planning some terrorist attack on the basis of intelligence investigations. On the other hand, for this same reason, in order to be morally justified, a killing in self-defence should not be planned and programmed but should rather be the outcome of a last resort attempt of incapacitating a threatening aggressor; in this respect, targeted killings by drones seem particularly problematic because armed drones – like traditional missiles and unlike guns – are designed not to capture or incapacitate but only to kill.

Moreover, permissive positions on the use of drones for extraterritorial targeted killings like Kasher's can be challenged also from the perspective of international law. Kasher's reliance on a permissive interpretation of the doctrine of self-defence as a justification for targeted killings by drones seem to leave too much space for states to elude other of their responsibilities. Firstly, whereas self-defence may sometimes make some killings all things considered justified, self-defence is not a licence to kill, that is, a full exemption from responsibility. This means, in particular, that those who kill in self-defence are still accountable for what they did; that is, they have to give a full report of their behaviour, not least in order to give evidence that the conditions for lethal self-defence were met in that specific case. But Chantal Meloni claims in Chapter 2 that, as current lethal drone attacks are made on the basis of undisclosed intelligence information and reports, states are de facto not accountable for their extraterritorial killings by drones; in the absence of a reliable account of facts, we run the serious risks of leaving serious war crimes unpunished. And in any case, the use of drones for targeted killings may create "a major accountability vacuum" (Alston 2010).

Secondly and relatedly, a justification does eliminate blameworthiness, but it does not eliminate causal responsibility and the related duty to apologize to and possibly compensate the victims of one's actions. However, when states are not held accountable for their operation and they don't act according to their duty to investigate the effects of their operations, there may be no official way to ascertain the number and identity of victims – be they the intended targets of the attack or the unintended side effect of it (i.e. innocent civilians), so that, even assuming that both the intended and the unintended killings were justified (the former by the doctrine of self-defence, the latter by the doctrine of necessity and proportionality) – there would be no way for the state to discharge its duty to apologize and maybe compensate those victims.

Finally, states also have a duty to protect their soldiers' lives and health. Indeed, one common point of discussion in the debate on the permissibility of the use of military drones is whether it is all right for a state to "kill by remote control" (Strawser 2013). Supporters of drones insist that drones help states fulfil their duty to keep soldiers as removed as possible from the dangers of the battlefield and, more generally, provided certain requirement are met, to arguably improve the overall well-being of people (Müller, Chapter 3 in this volume). Critics observe that the remoteness from the battlefield and the related safety for military personnel that military drones allow may be ethically problematic. It may create an unjustifiable asymmetry in war between actors that possess drones and actors who do not (Enemark 2014; Galliott 2012; Kahn 2002) and/or a dangerous lightheartedness among the military personnel (the "Playstation mindset" feared, among others, by Alston 2010) or a lowering of the threshold for waging war by states. Without denying the obvious moral advantages stressed by drone supporters, Koch (Chapter 4 in this volume) goes as far as to claim that this march towards more and more safety for soldiers may also be seen as intrinsically wrong in some respect. According to Koch's "existentialist" perspective, the state's responsibility to grant its citizens and soldiers as much safety and well-being as possible may be in conflict with the individuals' responsibility towards themselves, that is, people's duty to be faithful to their moral identity; for soldiers, Koch argues, this identity critically depends on their putting their lives at risk.

As distant as they may be in their moral and political evaluations, what both parties in this debate seem to agree on is that drone war is not dangerous for drone operators. However, as extensively argued in Chapter 5 by Jesse Kirkpatrick, this may not be the case. Drone combat may actually share important features with traditional combat, and drones operators may be thus exposed at least to a similar psychological harm as traditional soldiers are. If this is the case, states may also have a duty to recognize drone operators as professional combatants, not just lighthearted computer geeks, and to make sure that drone operators have all the kinds of support they may need (e.g. adequate training, medical and psychological help, etc.).

5.2 Individual Responsibilities of Military Personnel and Politicians

A second cluster of concerns relates to the moral and legal responsibilities of individual state officials and military personnel who deploy drones in lethal operations. By examining the legal cases following the targeted killings of Al-Aulaqi by the U.S. government and of Salah Shehadeh by the Israeli government, Meloni highlights some difficulties in the ascertainment of legal responsibility before domestic courts for targeted killings by drones and their collateral effects. However, things may be different with the use of drones in more conventional war operations.

In his detailed legal Chapter 1, Dan Saxon – a former prosecutor at the International Criminal Tribunal for the former Yugoslavia – analyzes how the process of assessment of criminal responsibility for war crimes may be affected by the use of military drones. Saxon analyses two kinds of legal responsibility: (1) the liability of individuals who directly support with their acts the commission of violations of the laws of war by the use of drones and (2) the "superior" or "command" responsibility deriving from the failure of military and civilian superiors to perform their duty to prevent their subordinates from committing such crimes (a typical example of role-responsibility). Saxon admits that the presence on the battlefield of complex automated or semiautomated weapon systems like drones may complicate the task of commander to supervise and make sure that no mistakes are done in war operations and that an "accountability gap" would exist in circumstances in which a drone makes a "mistake" leading to an unlawful attack, while at the same time the commander could not in any way anticipate that this could happen. However, Saxon claims, this "would not be qualitatively different from other 'accountability gaps' found in modern warfare and tolerated by international law" (Saxon, this volume). Moreover, the use of military drones may sometimes even enhance the capacity to trace legal responsibilities due to the ability of digital technology to record and store information about its activity.

Legal considerations apart, one may still wonder whether the introduction of a (semi)automated machine mediating warfare operations may sometimes offer individual military personnel a valid excuse from moral responsibility in relation to some unlawful behaviour. Since World War II and the ensuing Nuremberg Trials, soldiers responsible for the commission of war crimes are not allowed any more to defend themselves by appealing to the fact that they followed orders. However, in Chapter 9, Alex Leveringhaus wonders whether in the future, drone operators will be able to legitimately appeal to the fact that they followed "what automated mechanisms told" them. For instance, an unlawful killing of innocent civilians may be caused by the operator's relying on what turned out to be wrong information provided by the automatic observation system of the drone. In such a situation, it seems that the operator may point to his ignorance (or, in Leveringhaus's words, lack of moral awareness) as a possible excuse for his misbehaviour; in fact, he may successfully argue that he simply didn't know that there were innocent civilians in the targeted area. However, Leveringhaus argues, whereas the operator's behaviour may be judged as non-culpable by focusing on the circumstances at the time of the deployment of the drone, he can still can be judged negligent for deploying and relying on a technology that put innocent lives at risk (under certain circumstances) or even for simply not taking "adequate steps to attain the morally relevant facts that enabled him to assess the risks arising from automation" (Leveringhaus, this volume).

In Chapter 11, on the relationship between drones and moral responsibility, Nikil Mukerji, while dissenting on Sparrow's thesis that the introduction of autonomous robots would lead to a responsibility gap, agrees that there should be limits to the autonomy of military action granted to drones; by adapting his terminology to that used in this introduction, his main argument is that machines that are not responsible in the capacity sense – that is, are not responsive to the variety of reasons for action typically responded to by human agents – should not be granted the possibility to take morally relevant decisions – in this case, killing – no matter how sophisticated their capacity for perception, calculus, and the like may be.

However, in his Chapter 6, Asa Kasher disagrees with a similar a priori moral approach and claims that military drones may sometimes be allowed to take life-and-death decisions, provided they pass a variation on Alan Turing's famous Imitation Game, that is, provided they can complete certain (portions of) military tasks in a way that is behaviourally indistinguishable from the performance of a morally and legally competent human being.

Finally, in Chapter 10, Michael Funk, Bernhard Irrgang, and Silvio Leuteritz reflect on the relationship between human autonomy and the development of (military) autonomous technology from a broader philosophical and anthropological perspective. Their normative conclusion is that, in contrast with their tendency to mirror themselves in their technological creations, humans should remain aware of the specificity of their autonomy of action and not give up their political responsibility to regulate the use of these technologies in the light of ambitious moral ideals.

5.3 Designers' and Programmers' Responsibilities

Leveringhaus claims that not only drone pilots but also drone programmers may be legitimately burdened with a high demand for moral awareness because of the high stakes involved in the use of automated weapon systems and because of their being in the position of taking reflective choices far away from the heat of the battlefield.

This approach is consistent with the so called value-sensitive design (VSD) approach in engineering. Scientists and engineers are very often causally responsible for serious political and military catastrophes. To mention one standard example, the atomic bombing of Hiroshima and Nagasaki would not have been possible without the research on nuclear energy conducted by outstanding scientists and engineers. It is debatable whether and under which conditions scientists can be held morally – or even legally – responsible for the unlawful use of the technology that they contributed to create. Arguably, if scientists and engineers were aware of the possible catastrophic misuses of these technology, they may share some moral responsibility for these misuses.

However, according to the value-sensitive design approach, the main focus of the ethical reflection on technology and responsibility should not be on the issue of backward-looking moral responsibility of engineers and designers but rather on the question about their forward-looking moral responsibilities. In this perspective, designers and programmers should anticipate possible ethical, political, and societal issues and risks posed by their technology, and they should try to design technology, from the early stages, in such a way as to address these issues and to minimize those risks. From this perspective, Tjerk de Greef (Chapter 7) argues that the challenge for a drone designer is finding a way to avoid accountability gaps. Automated machines are not morally responsible – humans operators are. Yet designers are responsible too; they ought to make sure that operators relying on automated systems not only receive the best available information from the system; they should also be able to assess the reliability of this information themselves. In order to achieve this goal, machines have to be designed and programmed in such a way as to become, in a metaphorical sense, "accountable" partners to their operators. Operators must be put in the position to understand the "reasoning" behind a certain bit of information or suggestion provided by the machine and to reach their final decision by actively interacting with the machine.

While sharing a value-sensitive design approach to military drones, Aimee van Wynsberghe and Michael Nagenborg (in Chapter 8) broaden its application beyond the limits of war operations. If one considers the history of the technology deployed in warfare in the past decades – so they argue – and in particular how technology designed for military purposes has become widely used in civilian contexts – the Internet is just one striking example – one must conclude that there is no such thing as a "military technology". But if this is the case, and drones will be used more and more in different civilian contexts (e.g. in agriculture, in policing, in rescue and disaster management), it is the designers' responsibility also to manage this transition by starting anticipating which values are at stake in these other domains as opposed to the military domain and by working on how to redesign drones in such a way as to respect and promote these other values.

Note

1 Kasher explicitly applied this reasoning to targeted killings *by drones* in his presentation at our Robowar Responsibility workshop, Delft Univeristy of Technology, 28–29 March 2014.

References

Alston, Philip. 2010. *Study on Targeted Killing, Addendum to the Report of the Special Rapporteur on Extrajudicial, Summary or Arbitrary Executions*, 28 May, A/HRC/14/24/Add.6 (Alston 2010 Report)/. Available online at http://www2.ohchr.org/english/bodies/hrcouncil/docs/14session/A.HRC.14.24.Add6.pdf.

Enemark, C. 2014. *Armed Drones and the Ethics of War*. Routledge.

Galliott, JC. 2012. Uninhabited Aerial Vehicles and the Asymmetry Objection: A Response to Strawser. *Journal of Military Ethics* 11(1): 58–66.

Hart, HLA. 1968. *Punishment and Responsibility*. Oxford University Press.

Kahn, PW. 2002. The Paradox of Riskless Warfare. *Philosophy and Public Policy Quarterly* 22(2): 2–9.

Kasher, A. 2014. The Ethics of Protective Edge. *The Jewish Review of Books* Fall.

Matthias, A., 2004. The Responsibility Gap. Ascribing Responsibility for the Actions of Learning Automata. *Ethics and Information Technology* (6): 175–183.

Sparrow, R. 2007. Killer Robots. *Journal of Applied Philosophy* 24(1): 62–77.

Strawser, BJ. 2013. *Killing by Remote Control: The Ethics of an Unmanned Military*. Oxford University Press.

van de Poel, I., Royakkers, L. and Zwart, SD. 2015. *Moral Responsibility and the Problem of Many Hands*. Routledge.

Vincent, NA. 2011. A Structured Taxonomy of Responsibility Concepts, in N Vincent, I van de Poel & J van den Hoven (eds). *Moral Responsibility: Beyond Free Will & Determinism*. Springer, 15–36.

Part I

Drones and Legal Responsibility

1 Autonomous Drones and Individual Criminal Responsibility

Dan Saxon

1. Introduction

This chapter will describe the impact that the introduction of autonomous drones to the battlespace may bring to efforts to hold individuals responsible for violations of the laws of war and gross violations of International Human Rights Law, such as crimes against humanity.[1] It will review theories of criminal responsibility that may be applicable when autonomous drones are employed to perpetrate crimes. I argue that the use of kinetic autonomous drones should not necessarily produce unacceptable 'accountability gaps.' However, an 'accountability dichotomy' may develop between states with the resources to field autonomous drones that can communicate and record all relevant military data and less sophisticated states and armed groups that do not have access to this technology. Moreover, as the speed and military advantage of autonomous technology increases, a danger exists that drone operators and their commanders will gradually cede their human control and supervision over such weapons. In that scenario, proving individual criminal responsibility for incidents involving autonomous drones will be more challenging. For the purposes of this chapter, the term 'autonomous drones' refers to airborne weapon platforms that have the capability to make and execute targeting decisions without human supervision (U.S. Department of Defense 2012: 13–14).

1.1 Autonomous Drones and Challenges for Commanders

Under international law, the existence of individual criminal responsibility often turns on the intent or *mens rea* of the accused. Deployment of an autonomous weapon system, however, suggests a far more complex mental process, than, for example, firing a pistol at the enemy or the deployment of a unit of human-operated drones in the battlespace. The soldier who aims his pistol at a human target (1) will understand the capability of that weapon to injure or kill the target and (2) will 'know,' to the best of her ability, whether the target is an enemy combatant or a civilian. Similarly, the General who deploys a fleet of drones into a battlespace must consider, *inter alia*, (1) the range, accuracy, and explosive power of the drone's weaponry to be directed at the enemy, (2) the possible presence of civilians in the area, and (3) the levels of training, stamina, discipline, and supervision of the drone's human pilots, including their ability to comply with International Humanitarian Law (IHL).[2] Thus, the General who deploys human drone crews must understand and have confidence in the ability of those soldiers to think about the changing circumstances of their battlespace and to make appropriate judgments consistent with their mission.

Let us assume, however, that in 2065 the General deploys a brigade of autonomous drones. These robotic airborne vehicles, possessing great firepower, would have sufficient artificial intelligence to locate and identify targets that have been programmed into their computer software and destroy those targets, while respecting the rules of IHL and any pertinent rules of engagement. The option to deploy unmanned aircraft without human pilots and gunners on board could save many lives.

Nevertheless, this option also would increase the demands on the General's already taxed mental capacity. Prior to such deployment, the General must consider (1) the first two points previously mentioned vis-à-vis human-operated drones, (2) the autonomous drone's ability to comply with IHL in the particular (fluid or not) battlespace, (3) whether the mission or the expected circumstances of the battlespace may require the exercise of increased levels of human supervision and control over the robotic airborne vehicles (i.e. restrictions on autonomy), (4) whether the General and/or her staff will have the capacity to deactivate immediately the autonomous drone should conditions require it, (5) the robustness of the software that operates the artificial intelligence of the autonomous drone, in particular whether enemy forces may have the ability to tamper with and/or take control over the autonomous drone(s), (6) the level of training – technical, operational and with respect to the laws of war – of the human 'operators' or monitors of the autonomous weapon systems (if any), and (7) whether the General and/or her staff will be able to monitor the rapid communications that will occur *between* autonomous drones and intervene when necessary. This last challenge will be particularly complex if the autonomous drones are deployed as a form of 'swarm' technology, i.e. large numbers of autonomous mobile weapon platforms that communicate amongst themselves and adjust their behaviour – individually and as a 'swarm' – when necessary to achieve the mission.[3]

Should one or more of the autonomous drones attack and kill civilians, on what basis might individual criminal responsibility accrue to the General who ordered their deployment? If the General intentionally deployed the autonomous drones with the knowledge that they were deficient vis-à-vis one or more of the targeting rules mentioned in Articles 48–59 of Additional Protocol I (Geneva Conventions Additional Protocol I 1977) or with deliberate indifference to the existence of such deficiencies, arguably the General conducted an indiscriminate attack for which she should be held accountable [Geneva Conventions Additional Protocol I 1977: Art. 51(4), Art. 85(3)(b)].

From a legal perspective, more complex tasks and/or more limited autonomous technology will signal a demand for greater levels of human judgement (as well as communication and accountability) during the mission. Therefore, a sound understanding of the function, capabilities, and limitations of the semi-autonomous and autonomous weapon technologies available to armed forces will become a prerequisite for the command of modern military units. Increased autonomy will enable higher speeds of military engagement, resulting in greater military advantages for the armies that field them. In parallel, considerations of military advantage mean that the increasing speed with which future weapon systems will absorb, process and transmit information and react to events may influence decisions about acceptable levels of human judgement and permissible levels of autonomy (Jackson 2014).

1.2 Implications of the Inevitable Employment of Increasingly Autonomous Drones

The potential military advantages for states and organized armed groups in possession of autonomous weapon technologies and capabilities will lead, inevitably, to the increasing use of autonomous drones (Corn 2014). However, to ensure compliance with IHL, the intervention of human judgement on the activity of autonomous drones will be required at four distinct stages of military operations: (1) the procurement/acquisition stage, (2) the planning stage of the mission or attack when a human must choose which weapon system to employ (systems will vary across a range of autonomy), (3) following the choice of an autonomous drone, a decision as to the level of human attention – if any – to assign to the system during the mission but prior to an attack, and (4) specific inputs of human judgement – if necessary – to comply with international legal obligations and/or political interests, immediately before, during, and after the attack. During phases 2–4, human control over the autonomous drone should remain until the human supervisor ensures, at each of the final three stages, that the weapon system complies and will continue to comply with international law and applicable rules of engagement (Saxon 2014).

When accidents or crimes occur during and/or due to the use of autonomous drones, the actual degree of autonomy and the levels of human supervisory control over the machine must form part of any accountability analysis. Once the speed of autonomous technology reaches levels that preclude effective human supervision and control, however, proof of the existence of the *mens rea*[4] may be illusory and/or impossible to establish. Arguably, this could result in an accountability gap as the underlying rationale for the *mens rea* requirement in criminal law is that a sense of personal blame is absent if the accused did not in some way intend her action or omission (Supreme Court of Canada *R v. Finta* 1994: 760).

2. Applicable Law

2.1 Targeting Rules and Unlawful Attacks under IHL

The rules governing the selection and attack of targets in Articles 48–59 of API will govern many of the crimes where autonomous drones are employed. Articles

48 and 52 enshrine the duty of parties to an armed conflict to distinguish between the civilian population and combatants and between civilian objects and military objectives and thus to direct operations only against combatants and/or military objectives.[5] Article 51(4) prohibits indiscriminate attacks, which include:

(a) those that are not directed at a specific military objective;[6]
(b) those that employ a method or means of combat that cannot be directed at a specific military objective; or
(c) those that employ a method or means of combat the effects of which cannot be limited as required by API.[7]

Article 54, which reflects customary international law [Schmitt 2013: Rule 81 (1)], prohibits attacks against objects that are indispensable to the survival of the civilian population 'for the specific purpose of denying them for their sustenance value to the civilian population or to the adverse Party, whatever the motive'. Such indispensable objects would include food supplies, crops ripe for harvest, drinking water reservoirs and water distribution systems.[8] Article 56 bans attacks against works or installations containing dangerous forces, i.e. dams, dykes, and nuclear power plants.

Article 57 addresses the required precautions that 'those who plan or decide upon' an attack must exercise due diligence to avoid or minimize civilian casualties. Planners and executors of attacks must do everything feasible to verify that the target of the attack is a military objective, and the provisions of the Additional Protocol I (API) do not forbid the operation [Geneva Convention API: Art. 57(2)(a)(i)].[9] Preliminarily, as Boothby observes, Article 57 contains no requirement that persons execute attacks (Boothby 2012: 20). However, in 1977 when the drafters of API completed their work, autonomous weapon systems were not foreseeable. Indeed, during the drafting process of Article 57, the 'relevant discussion' as to the meaning of 'those who plan or decide upon an attack' concerned the question whether Article 57 obligations could be fulfilled only at a certain level of command, given that planners and decision-makers need appropriate information on which to base considerations about precautionary measures.[10] Thus, it would be inaccurate to suggest that the drafters of Article 57 intended the scope of the word 'those" to encompass autonomous machines.[11] IHL must apply to the use of autonomous weapon systems (like any other weapon systems). Therefore, the onus for taking the precautionary measures described in Article 57, therefore, must remain with the human commanders and operators who have the capacity to exercise their judgement over the employment of autonomous weapon systems.

In addition, belligerent forces must 'take all feasible precautions in the choice of means and methods of attack' [Geneva Convention API 1977: Art. 57(2)(a)(ii)] to avoid and minimize incidental injury to civilians and damage to civilian objects.[12] "Feasible precautions" are precautions that are practicable or practically possible considering all circumstances ruling at the time, including humanitarian and military considerations (UN CCW Protocol II: Art. 10). Thus, this duty does not require an attacker to be certain that the target of the attack is lawful (Dinstein

2010: 139; Boothby 2012: 121). Instead, the obligation is to act with due diligence and in good faith (Dinstein 2010: 139).

Given the complexities of combat and the battlespace, the general principles of Article 57 do not give rise to specific rules that particular types of weapons must be used in a specific case. Instead, parties to armed conflict 'retain considerable discretion, which will be exercised primarily according to military considerations, the framework of operational priorities, and not mainly according to humanitarian aspects' (Oeter 2008: 189–190). Nevertheless, given the relative 'brittleness' of present-day autonomous weapon systems, for the foreseeable future, the duty to take 'feasible precautionary measures' during the planning and execution of attacks may require human supervision – at the expense of speed and military advantage – over the final determination as to whether a target is lawful.

An important value underlying the rules concerning precautionary measures and a fundamental principle of IHL is that civilians and civilian objects shall be spared as much as possible from the effects of hostilities (ICTY *Galić* Appeal: para. 190). Accordingly, the rule of proportionality, expressed in API Articles 51(5)(b) and 57(2)(a)(iii), requires that parties to armed conflict 'refrain from deciding to launch any attack which may be expected to cause incidental loss of civilian life, injury to civilians, damage to civilian objects, or a combination thereof, which would be excessive in relation to the concrete and direct military advantage anticipated.'[13] The International Committee of the Red Cross (ICRC) *Commentary to Article* 52(2)(a)(iii) observes that 'a military advantage can only consist in ground gained and in annihilating or weakening the enemy armed forces' (ICRC *Commentary*: para. 2218). Nevertheless, 'an attack' in this context may be comprised of a number of coordinated actions. Thus, the military advantage anticipated from an attack refers to the advantage expected from the attack considered as a whole and not only from isolated or specific parts of the attack (U.K. Joint Doctrine and Concepts Centre 2004: para. 5.33.5).

Proportionality analyses, often made during the stress of military operations, are notoriously difficult and involve a degree of subjectivity. In 2000, a report issued by the International Criminal Tribunal for the former Yugoslavia (ICTY), observed that '[i]t is much easier to formulate the principle of proportionality in general terms than it is to apply it to a particular set of circumstances because the comparison is often between unlike quantities and values. One cannot easily assess the value of innocent human lives as opposed to capturing a particular military objective' (ICTY *Final Report*: para. 48).

The ICRC *Commentary* suggests that the rule, 'such as it is' (ICRC *Commentary*: para. 2219), attempts to *balance* the competing interests of military necessity and the protection of civilian populations.[14] However, the ICRC *Commentary* also states that '[i]ncidental losses and damages should *never* be extensive' (ICRC *Commentary*: para. 1980, emphasis added). More recently, a group of IHL experts, in disagreement with the later approach, opined that 'extensive collateral damage may be legal if the anticipated concrete and direct military advantage is sufficiently great. Conversely, even slight damage may be unlawful if the military advantage expected is negligible' [Schmitt 2013: Rule 51(7)]. An ICTY Trial Chamber adopted a standard that (apparently) would not criminalise extensive collateral damage as long as the damage was not excessive in relation to the expected military advantage.[15]

Finally, Article 59 prohibits attacks 'by any means whatsoever' against non-defended localities.[16] Conversely, attacks against defended positions are not forbidden (Dinstein 2010: 108).

Although not all states are parties to the four Geneva Conventions of 1949 and the Geneva Conventions' Additional Protocols I (1977), many IHL treaty rules reflect customary International Humanitarian Law.[17] To anchor individual responsibility for serious violations of international law in customary law, there must be state practice and *opinio juris* (Simma and Paulus 1999: 309).[18] In the context of IHL, where multiple treaties codify legal obligations during armed conflict, the primary significance of a norm's customary character is that the norm binds states that are not parties to the instrument that restates the norm (Meron 1989: 3).

IHL rules and principles concerning means and methods of warfare during international armed conflicts have gradually extended to civil wars (ICTY *Tadić*: para. 119). Thus, customary international law criminalizes violations of the laws or customs of war, whether committed in international or non-international armed conflicts (ICTY *Galić* Appeal: para. 120).

The pre-existence of law criminalizing the improper use of autonomous weapons systems will be essential to any efforts to enforce individual criminal responsibility. Articles 65 and 67 of Geneva Convention IV of 1949 (Relative to the Protection of Civilians in Time of War) enshrine in IHL the fundamental legal principle that penal laws cannot be retroactive. Thus, for example, the courts of occupying powers may apply only those laws that were in force at the time an alleged criminal act was committed. This norm of criminal law (*Nullum crimen, nulla poena sine lege*) reduces the possibility of arbitrary actions by courts and ensures that 'penal jurisdiction is exercised on a sound basis of universally recognized legal principles' (ICRC *Commentary*: Geneva Convention IV, Art. 67).[19]

The fact that autonomous weapon systems are 'new' weapons, however, should not give rise to a *nullum crimen* problem. The question for criminal courts is whether evidence proves that a human employed the weapon unlawfully. Nor should the fact that the computer software controlling the autonomous weapon system makes the final decision(s) concerning attacks necessarily raise questions of retroactivity. If evidence establishes that human commanders or operators employed the autonomous weapon system with the intent to carry out an unlawful attack and the unlawful attack occurred, responsibility will be similar to that of a commander who plans or orders attacks with artillery that violate IHL. In that scenario, there is no dispute that both customary and treaty law provide for individual criminal responsibility in the case of unlawful attacks [ICTY *Galić*: para. 130; ICTY *Strugar*: para. 275; Henckaerts and Doswald-Beck (2005: 25–71, 551–563);[20] Rules 7–22 and 151–153].

2.2 The Law of Individual Criminal Responsibility

In circumstances of armed conflict, two kinds of individual criminal responsibility may arise when soldiers and/or their commanders violate the legal rules just described and perpetrate unlawful attacks as described.[21] First, 'direct' responsibility arises from an individual's acts or omissions that support the commission of violations of the laws of war (ICTY *Galić*: para. 169). Second, 'superior' or 'command' responsibility emanates from the failure of military or civilian superiors to perform their duty to prevent their subordinates from committing such crimes and/or the failure to fulfil the obligation to punish the perpetrators thereafter (ICTY *Perisić*: paras. 86–87).[22]

In situations outside of armed conflict, where autonomous drones are used, for example, to perpetrate crimes against humanity, International Human Rights Law and international criminal law would provide the relevant legal framework. Evidence that a state or organized group employed autonomous weapons to conduct widespread or systematic attacks against a civilian population would be required.[23] However, the general theories of individual criminal responsibility would remain the same.

2.2.1 Direct Responsibility

The use of autonomous drones for the perpetration of crimes most probably will involve one of five modes of direct responsibility: commission, planning, ordering, aiding and abetting, or attempt. For the purposes of brevity, this chapter will focus on the theories of commission, planning and ordering.

Commission of a crime entails the physical perpetration of a crime or engendering a culpable omission in violation of criminal law (ICTY *Limaj*: para. 509). The *actus reus* of this mode of criminal liability is that the accused participated, physically or otherwise directly, in the material elements of a crime, through positive acts or omissions, whether individually or jointly with others (Ibid.: para. 509). In addition to the *actus reus*, as mentioned, attribution of criminal responsibility for any of the crimes that fall within the jurisdiction of the ad hoc tribunals and/or the International Criminal Court (ICC) 'depends on the existence of the relevant state of mind or degree of fault.'[24] Thus, the requisite *mens rea* for commission is that the perpetrator acted with the intent to commit the crime, or with an awareness of the probability, in the sense of the substantial likelihood, that the crime would occur as a consequence of his/her conduct (ICTY *Limaj*: para. 509).[25]

In addition, at the ICC, criminal responsibility may accrue when accused make an essential contribution to a plurality of persons acting with a common criminal purpose. The accused must be aware of her essential contribution and must act with the intention that the crime occur or with the awareness that by implementing the common plan, the crime 'will occur in the ordinary course of events.'[26] At the ad hoc tribunals, culpable participation in a common criminal purpose is referred to as 'joint criminal enterprise' and requires a significant contribution to the realization of the crime (ICTY *Popović* et al.: para. 1027, citing ICTY *Krajišnik*: para. 215; ICTY *Brđanin*: para. 430; ICTY *Kvočka* et al.: para. 97).

The *actus reus* of *planning* requires that one or more persons design the criminal conduct constituting one or more statutory crimes that are later perpetrated (ICTR *Nahimana*: para. 479). It is sufficient to demonstrate that the planning was a factor substantially contributing to such criminal conduct (ICTR *Nahimana*:

para. 479). The *mens rea* for this mode of responsibility is the intent to plan the commission of a crime or, at a minimum, the awareness of a substantial likelihood that a crime will be committed in the execution of the acts or omissions planned (ICTR *Nahimana*: para. 479).[27]

Responsibility under the mode of *ordering* ensues when a person in a position of authority orders an act or omission with the awareness of the substantial likelihood that a crime will be committed in execution of that order and if the person receiving the order subsequently commits the crime. Orders need not take a particular form, and the existence of orders may be established using circumstantial evidence. Liability ensues if the evidence demonstrates that the order substantially contributed to the perpetrator's criminal conduct (ICTR *Nahimana*: para. 481; ITCY *Boškoski & Tarčulovski*: para. 160).

2.2.2 Superior Responsibility

The theory of superior responsibility entails accountability based on a commander's omission rather than commission or participation in crimes. Evidentiary requirements to establish a commander's superior responsibility for crimes differ between the ad hoc international criminal tribunals and the ICC. We will first discuss the standards implemented by the ad hoc tribunals and then describe where the Rome Statute demands different levels of proof. At the ad hoc international criminal tribunals, a superior will be held accountable for crimes committed by subordinates if the prosecution demonstrates beyond a reasonable doubt:

- i The existence of a superior–subordinate relationship was demonstrated by the commander's 'effective control' over the persons who commit the crime;
- ii the superior knew or had reason to know that the criminal act was about to be or had been committed; and
- iii the superior failed to take the necessary and reasonable measures to prevent the criminal act or punish the perpetrator thereof (ITCY *Perisić*: para. 86).

(A) SUPERIOR–SUBORDINATE RELATIONSHIP

The superior–subordinate relationship lies at the heart of the doctrine of a commander's liability for crimes committed by her subordinates. The role of commanders is decisive,[28] and it is the position of command over and the power to control the acts of subordinates that form the legal basis for the superior's duty to act, and for her corollary liability for a failure to do so (ICTY *Limaj*: para. 76).

(B) THE SUPERIOR KNEW OR HAD REASON TO KNOW

A superior's knowledge that her subordinates were about to commit or had committed crimes may be actual knowledge or the availability of specific information that would put her on notice of these events (ICTY *Strugar*: paras. 302 and 304; ICTY *Limaj* et al.: para. 525).[29] The information available need not be sufficient, by itself, to compel the conclusion of the existence of such crimes (ICTY *Strugar*: para. 525). It is sufficient that the superior has possession of 'sufficiently alarming' information, even general in nature, putting a superior on notice of the risk of illegal acts by her subordinates, so as to justify further inquiry to ascertain whether crimes were actually being or about to be committed (ICTY *Strugar*: para. 525; ICC *Bemba*: para. 434).[30] Such knowledge may be presumed if the superior had the means to obtain the knowledge but deliberately refrained from doing so (ICTY *Delalić*: 226). Importantly for the purposes of this chapter, a superior's failure to remain apprised of her subordinates' conduct or to establish a monitoring system may constitute a lapse of duty that results in responsibility within the military disciplinary framework. It will not necessarily, however, result in criminal liability (ICTY *Delalić*: 226).

(C) FAILURE TO TAKE NECESSARY AND REASONABLE MEASURES TO PREVENT OR PUNISH

With respect to the third element of superior responsibility, the assessment of 'necessary and reasonable' measures to fulfil a commander's duty is not a matter of substantive law but of evidence (ICTY *Halilović*: para. 63). 'Necessary' measures are the measures appropriate for the superior to discharge her obligation (showing that she genuinely tried to prevent or punish), and 'reasonable' measures are those reasonably falling within the material powers of the superior (ICTY *Halilović*: para. 63). A superior will be held responsible if she failed to take such measures that are within her material ability and the superior's explicit legal capacity to do so is immaterial provided that she had the material ability to act (ICTY *Limaj*: para. 521; ICTY *Aleksovski*: para. 521; ICC *Bemba*: para. 443).[31]

The Rome Statute of the ICC alters the evidentiary thresholds for holding civilian and military commanders accountable under the theory of superior responsibility.[32] In addition to the three elements found in the law of the ad hoc tribunals, prosecutors at the ICC must establish that the crimes committed by subordinates occurred as a result of the superior's 'failure to exercise control properly over such forces' [ICCSt: Art. 28(1)]. In short, it is necessary to prove that the superior's omission increased the risk of the commission of the crimes charged (ICC *Bemba*: para. 425).[33]

Moreover, instead of requiring proof that the superior 'had reason to know' that her forces were committing or had committed crimes, the ICC 'knowledge' standard for military commanders compels prosecutors to establish that she 'should have known' about such crimes.[34] This standard requires the commander 'to ha[ve] merely been negligent in failing to acquire knowledge' of her subordinates' unlawful conduct.[35] The 'knowledge' requirement for demonstrating liability of civilian superiors is higher: 'the superior either knew, *or consciously disregarded* information which clearly indicated that the subordinates were committing or about to commit such crimes.'[36]

The 'failure to punish' prong common to the law of the ad hoc tribunals is replaced at the ICC by a 'failure to repress" element, which includes two distinct duties arising at different stages of criminal conduct. First, the duty to repress includes the obligation to stop ongoing crimes.[37] Second, the duty to repress encompasses the obligation to punish subordinates after the commission of crimes.[38] Finally, at the ICC, superiors may also be held accountable specifically for failing 'to submit the matter to the competent authorities for investigation and prosecution.'[39]

3. Application of the Law of Individual Criminal Responsibility to the Use of Autonomous Drones

Prosecution of military or civilian superiors for the unlawful use of autonomous drones would require proof of the unlawful attack and the alleged mode of individual criminal responsibility. In order to demonstrate violations of particular targeting rules in Articles 48–59 of API, courts would need evidence of, *inter alia*: the protected status of victims (ICTY *Boskoski & Tarčulovski*: paras. 51, 134, 138–140, 167, 172, 310–311, 313–320, 326–328, 334–345, 383, and 388), the absence of military objectives in the area subject to attack,[40] the indiscriminate and/or disproportionate nature of an attack,[41] including the exercise or lack of precautionary measures taken by commanders planning and executing the attack,[42] and so on.

3.1 Direct Theories of Responsibility for Unlawful Attacks

In many cases involving unlawful attacks with the use of autonomous drones, proof of a commander's individual criminal responsibility under the direct modes of commission, planning and ordering could be less complex. If a tactical commander[43] intentionally employs an autonomous drone in circumstances (such as a densely populated urban area where civilians are known to be present) where the system's capabilities for IHL compliance are inadequate and deaths and injuries to civilians occur, that commander may be found culpable for the commission of a war crime.[44] Similarly, if a tactical commander orders the use of an autonomous drone in similar circumstances, and damage to civilian objects or injuries to civilians subsequently occurs, liability under this theory of responsibility could ensue. The same scenario should apply when commanders at the operational or strategic level deliberately order their subordinate units to employ particular autonomous drones in circumstances that are inappropriate for the expected capability of the weapons, resulting in civilian deaths and/or damage to civilian objects.

Although the location of proof – particularly of *mens rea* – always represents a challenge for prosecutors, one advantage of the sophisticated technology found in autonomous weapon systems is that the systems can be designed to leave an electronic 'footprint' of important decisions (Sassóli 2012: 316, 338). Indeed, at the development/procurement stage of autonomous drone technology, one requirement for validation of compliance with IHL could be a subsystem of electronic recording of commanders' operational and targeting decisions, such as (1) identity of target(s), (2) anticipated location of target, (3) anticipated civilian injuries,

(4) anticipated damage to civilian objects, (5) description of anticipated military advantage offered by neutralization of the target, (6) precautionary measures taken to avoid civilian injuries and damage to civilian objects, (7) foreseeable capacities of the autonomous weapon system selected for use, and so on.[45] In addition to the evidentiary trail produced by such electronic records, the creation of an obligatory record of decisions and evaluation of IHL duties – and the possibility of accountability – might focus the minds of field commanders so as to avoid criminal conduct.[46]

The availability of electronic records would also effectively minimize the challenge of physical and/or temporal 'remoteness.' Physical 'remoteness' in this context refers to the geographical distance between the acts or omissions of the superior and the location of the criminal conduct.[47] Temporal 'remoteness' refers to the time elapsed between the superior's acts or omissions and the execution of the crimes. Modern communications, however, increasingly provide superiors with real-time access to circumstances and events, including combat occurring far from command centers and headquarters.[48] Furthermore, the Internet and social-media technology create virtual links between front-line areas and all parts of the world. These connections, combined with records of commanders' decision-making processes, 'shorten' the physical and temporal distances between a superior's instructions and events in the battlespace. These relationships and interactions can reveal much about a superior's mental state when decisions affecting the use of force are made. Thus, evidence of the superior's intent and knowledge is far more relevant to individual criminal responsibility analysis than concerns about temporal or geographic 'remoteness.'[49]

Nevertheless, the development of 'moral remoteness' on account of increased interaction with, and reliance on, computerized systems may create real challenges to the establishment of criminal intent in military commanders. The tendency for human beings increasingly to depend on their computers at moments of decision-making can lead to 'moral buffering' and a reduced sense of responsibility for the consequences of those decisions. As artificial intelligence computer software for autonomous drones improves, a risk will accrue that commanders will shift the accountability for stressful decisions to the computer (Cummings 2006: 10–11, 18–19). In certain cases, that dynamic may obscure the presence of a criminal *mens rea*.

These hypothetical scenarios just described represent relatively clear-cut examples where direct criminal responsibility for employing autonomous drones to carry out unlawful attacks could be achieved. More complex issues arise, however, when the tactical commander conducts herself correctly in the selection and programming of an autonomous drone and potential targets, yet the weapon nevertheless attacks civilians or performs outside the confines of IHL. Given the lack of any criminal intent on her part, no criminal culpability should accrue to the field commander.[50]

Who, if anyone, should be held accountable in these situations? Professor Corn argues forcefully that, in these situations, the so-called Compliance Commander (i.e. the officer in charge of the procurement process of the new autonomous drone) should bear criminal responsibility for the misconduct of the autonomous weapon system (Corn 2014: 9–10 and 13). Standards for validation of new autonomous drones should be high enough to ensure their consistent functioning within their foreseen use and capacity. When the Compliance Commander approves the new weapon system, she takes responsibility for ensuring that these standards have been met and for failures that occur when the system does not operate according to these standards (Corn 2014: 9–10 and 13).[51]

Criminal responsibility for Compliance Commander may be appropriate if evidence exists that she intentionally and knowingly approved the procurement of a flawed autonomous drone or did so recklessly without first validating the systems capabilities, all possible failsafe mechanisms, all subsystems, and so on. However, if no such evidence exists, there would appear to be no basis – absent a 'strict criminal liability'[52] standard – for penal responsibility for the Compliance Commander.

Thus, without evidence of culpable *mens rea* for the field or Compliance Commander and absent a 'strict criminal liability' standard, an 'unlawful attack' perpetrated with the use of the autonomous drone would be a tragic accident and/or mistake and would be analogous to other kinds of military accidents where there is no human criminal responsibility.[53] Therefore, an 'accountability gap' would exist, but it would not be qualitatively different from other 'accountability gaps' found in modern warfare and tolerated by international law.[54] Furthermore, civil damages for accidents and/or mistakes involving non-autonomous weapon systems may not always be available.[55]

Multiple wrongs, of course, do not make right, and it is worth considering whether, in principle, even a limited accountability gap for kinetic autonomous drones is acceptable. Any gap in accountability could foster impunity and undermine faith in legal systems.[56] However, a gap might also – indirectly – stimulate technological development by reducing manufacturers' fears that they might be held responsible for mishaps or misdeeds involving autonomous drones. These realities illustrate accountability's tensions: too little may lead to injustice; too much may discourage legitimate scientific advancement.

Ironically, commentators raise concerns about accountability gaps for autonomous drones when we tolerate similar gaps for other kinds of complex weaponry. For example, accountability gaps are imaginable for the operation of complex non-autonomous weapon systems such as Tomahawk missiles, modern tanks, and jet fighter planes. One reason for this contradiction may be psychological and philosophical: the concept of autonomous weapons, for some persons, is disconcerting. Taken to its limit, it suggests an abdication of the most difficult decisions and actions of warfare (Heyns 2013: paras. 89–97).

That personal sense of disconcertedness, however, is irrelevant to the issue of whether accountability gaps are acceptable. The crucial question should be whether anticipated accountability gaps will be so onerous that they threaten the integrity of our legal system(s), in particular the requirements for individual criminal responsibility enshrined in the 'grave breaches' provisions of API and in customary law.[57] As long as persons who *intentionally* misuse autonomous drones can be held accountable, limited situations where individual criminal responsibility is absent should be permissible.

Moreover, as previously discussed, Article 57(2)(ii) of API requires commanders who launch an attack to 'take all feasible precautions' to minimize incidental injury to civilians and damage to civilian objects. Failure to comply with Article 57 obligations may constitute a grave breach and may be prosecuted as such (ICRC *Commentary API*: para. 2187). Boothby observes that while after-action inquiries may clarify whether the targets of attack are, or were, armed forces or civilians, 'any such determination during combat must usually be based on sketchy data and gut instinct' (Boothby 2012: 89). This perspective was generally true before the onset of the digital age, and even today, in the modern battlespace, amounts of electronic information are too overwhelming for humans to absorb. Conversely, future autonomous weapon systems will process this data (quickly) and respond according to the artificial intelligence software that controls its behaviour, without – potentially – the influence of human instincts. Thus, the introduction of autonomous drones adds a new layer to the discussion of what is 'practicable' in terms of feasible precautionary measures (Beard 2009: 434–349). That new dimension of technological power must be part of any analyses of potential criminal conduct involving the use of autonomous drones.

At first blush, an inherent contradiction appears to exist between the deployment of autonomous drones and Article 57(2). By definition, once the commander deploys an autonomous weapon platform, she may lose her ability to take (additional) feasible precautions.[58] Nevertheless, professional armies should continue to 'expect military commanders employing a system with autonomous functions to engage in the decision-making process that is required by IHL.' Thus, prior to deploying an autonomous drone, the commander must ensure one of two criteria: (1) once programmed, the artificial intelligence software controlling the autonomous drone has the robust capacity to determine and direct the feasible precautions necessary to comply with Article 57(2), or (2) deployment of the autonomous drone is itself an expression of a 'feasible precaution in the choice of means and methods of attack' within the meaning and spirit of the law (Jackson 2014).

These assessments will entail a degree of subjectivity.[59] They will also require, on the commander's part, a profound understanding of the functions and capacities of the artificial intelligence software controlling the available autonomous drones. That onus of understanding new and evolving technology is likely to be a heavy burden on the shoulders of future military and civilian commanders. It will require knowledge and comprehension of, *inter alia*, an autonomous weapon platform's on-board computer systems that make up its 'artificial intelligence', its navigational system, its communication abilities with humans and other autonomous systems, the kinds of weapons that it carries, and the capabilities of each weapon and its power/energy capacities. Logically, it would be impossible for commanders to *direct* weapons at specific military objectives, as required by Article 51(4)(b), without a proper understanding of the weapon. Thus, deployment of autonomous drones without a proper understanding of how the system works will constitute an indiscriminate attack and will be subject to criminal sanction,[60] at least in jurisdictions that recognize the *dolus eventualis* standard for *mens rea.*[61]

In addition, the commander will be obliged to ensure the technical knowledge of his subordinates who operate autonomous drones, as well as the subordinates' understanding of the laws of war.[62] Finally, the commander must monitor whether his subordinates deploy the autonomous drones appropriately, that is in accordance with IHL and any relevant rules of engagement.

Proof that a commander failed to inform herself of the functions and capacities of an autonomous weapon may come from several sources. One would be the education and training record of the commander; deployment of more complex weapons technology should require more sophisticated education and training. Another strategy would be to review the targeting protocol performed by the commander prior to launching the autonomous drone. For example, U.K. doctrine requires military planners to use a template called 'strategy to task to target' so that planners and commanders can demonstrate 'an audit trail, . . . from any target back to the strategy, thereby passing through all the aspects of the law of armed conflict such as discrimination, military necessity, et cetera' (Burridge 2009: 13). This record should include efforts by commanders in the field to seek the counsel and assistance of weapons experts prior to the deployment of an autonomous weapon system, that is links from 'the warfighter in the foxhole to the software engineer in the laboratory.'[63] Prosecutors, defence counsels, and judges also will have to familiarize themselves with autonomous weapon technologies in order to perform their tasks competently.

Similar analysis can occur when the use of autonomous drones results in alleged disproportionate attacks. As already noted, however, proportionality analysis requires the comparison of different values and usually entails a great deal of subjectivity. Not surprisingly, then, prosecutions of military commanders for conducting disproportionate attacks are notoriously difficult and relatively rare (ICTY Final Report).[64] As one commentator observes: 'on the accountability question, you read about the supposed "accountability gap" for killer robots. . . . But the "accountability gap" seems to have as a fiction behind it that every day we have prosecutions for the violation of the principle of proportionality. It's actually very difficult to hold someone accountable for violating the principle of proportionality in the battlefield' (Modirzadeh 2014). Thus, the use of autonomous drones during armed conflict would not likely give rise to a new 'accountability gap' for violations of the proportionality principle.

As autonomous weapon technologies develop, the use of early – relatively 'brittle' and less precise and reliable – autonomous drones will lead to an increased risk of individual criminal liability for the commanders who deploy them.[65] Hazards will remain, however, as algorithmic-based artificial intelligence evolves into 'machine learning'[66] and swarm technologies – large numbers of autonomous weapon systems that operate together and communicate with each other more rapidly than human commanders can monitor them – become a reality. Commanders will be challenged to (1) predict how 'learning' weapon systems will perform (to avoid accusations of launching indiscriminate attacks) or (2) control the rapid and complex actions of swarms.[67] There will be frequent tensions between a commander's exercise of her essential duty to command and control her subordinates and resources and her duty to accomplish her mission using a rapidly functioning weapon system that offers maximum military advantage and complies with IHL.

However, once the technology of autonomous drones passes a certain threshold for adaptability, reliability, and precision, the risk of individual criminal responsibility may be greater for commanders who *fail* to utilize them. As the president of the ICRC recently observed: '[i]n theory . . . it may be possible to program autonomous weapons to behave more ethically and cautiously on the battlefield than a human being. In such cases, the principle of precaution arguably imposes an obligation on States to choose the less harmful means to achieve their military aim, whenever feasible' (Maurer 2014:11). Thus, instead of producing an accountability gap, the development and deployment of autonomous drones from the arsenals of modern armies and armed groups may result in a 'accountability dichotomy' between states that have the resources to field the most sophisticated, reliable and accurate autonomous drones and those who employ more 'rudimentary' technology.

API does not explicitly oblige belligerent forces to employ the most sophisticated weaponry available to them. That textual construction may reflect the law of war's balance between the demands of military necessity and the principle of humanity. Nevertheless, future autonomous drones may well enjoy the advantages of speed and accuracy.[68] It would seem logical and not at all onerous to expect that a party in the possession of such weapons to use them whenever necessary to ensure all feasible precautions are considered and implemented.[69]

In her chapter in this volume, Chantal Meloni describes the Israeli government's failure to perform a criminal investigation of the 2002 bombing of the home of Salah Shehadeh, a Hamas military leader in the Gaza Strip. However, the Israeli government's failure to fully investigate that incident says nothing about the *ability* to investigate and prosecute persons responsible for war crimes that are committed using drone or other sophisticated technology. Any 'accountability vacuum' in such cases should not necessarily be attributed to the technology but rather to legal and/or political decisions not to pursue the matter.

3.2 Superior Responsibility

When crimes occur due to the misuse of autonomous weapons systems, the theory of superior responsibility may also be appropriate to hold commanders accountable. For example, if a commander at the operational level becomes aware that a subordinate officer at the tactical level is using autonomous drones to perpetrate unlawful attacks, the operational commander has a duty to prevent further misconduct and punish his subordinate (Geneva Convention API: Art. 86(2) and 87). As long as evidence exists demonstrating the three essential elements of a commander's effective control over subordinates who commit crimes – the commander's knowledge and a failure to take necessary and reasonable measures to prevent/repress further crimes and punish the perpetrators – criminal liability should ensue. As mentioned, instead of a soldier's or commander's 'direct participation' in a criminal event, the theory of superior responsibility imposes liability on the commander for her *omission*, for her failure to prevent further crimes and/or discipline those responsible.

The 'effective control' test to establish the existence of a superior–subordinate relationship should not differ when the principle 'perpetrators' of IHL violations employ autonomous machines to commit their crimes. However, the advances in technology may require reassessments of the 'knowledge' and 'reasonable measures' pillars of superior responsibility.

With respect to the knowledge element, commanders of forces operating autonomous drones will not easily convince a court that they were unaware that their subordinates were operating autonomous weapon systems unlawfully.[70] Any state or organized armed group with the resources and ability to field autonomous drones will also have the means and the communications technology to constantly monitor how these weapons are used. Furthermore, any competent commander would wish to utilize all possible methods to observe the progress and operations of her subordinate units.[71] Indeed, this full-time and real-time monitoring capability should be an obligatory aspect of every autonomous drone at the procurement stage (Trapp 2013: 156, 170).

However, the vast and overwhelming amounts of data 'available' to commanders will not guarantee a superior's *awareness* of particular information concerning the conduct of her subordinates. The ability of modern technology to acquire larger and larger amounts of data and/or obfuscate the reality of the battlespace may, at times, actually increase the fog of war (Dinstein 2010: 139). As Professor Cummings (a scientist and former fighter pilot) has observed, 'command and control technology have outpaced human reasoning capabilities and traditional command structures' (Cummings 2006: 17). Commanders should not, however, be permitted to use this challenge to abdicate their important role as insurers of subordinate compliance with IHL.[72] Thus, new systems of data capture, analysis, and human–machine interaction that prioritize information concerning the position, status, and conduct of subordinate forces – including the kinds of weapons employed by these units – will be necessary to ensure that commanders can effectively monitor the use of autonomous weapon systems (Cummings 2014).[73]

The theory of superior responsibility imposes a duty both to prevent the commission of the crime and to punish the perpetrators. The duty to prevent arises from the time a superior acquires knowledge or has reasons to know that a crime is being or is about to be committed, while the duty to punish arises after the superior acquires knowledge of the commission of the offence.[74] Thus, some level of interdependence – if not supervision – between commanders and the autonomous weapons systems that they employ will be necessary if the knowledge requirement for superior responsibility is to remain relevant.

The scope of necessary and reasonable measures to prevent further crimes and punish the subordinates involved in misconduct may vary when autonomous weapon systems are used to carry out unlawful attacks. For example, the use of swarm technology will undoubtedly increase the tempo of military engagements

(Fiddian 2012). The faster pace of combat – and unlawful conduct – will reduce a commander's opportunities to prevent crimes.

However, unlike human soldiers, many autonomous weapon systems can be switched off.[75] Thus, measures available to a commander to prevent or repress unlawful attacks carried out with autonomous drones should include efforts to electronically deactivate the machines. Indeed, part of the procurement process, including Article 36 legal reviews, should include a requirement for an 'override' mechanism permitting superiors to stop their subordinates' misuse of autonomous drones by taking control of or shutting down the machines. Use of override systems would constitute 'necessary and reasonable' measures to repress further unlawful conduct, and a commander's failure to make use of such mechanisms should result in liability.

Nevertheless, as Professor Heyns has observed, 'the power to override may in reality be limited because the decision-making processes of robots are often measured in nanoseconds and the information basis of those decisions may not be practically accessible to the superviser. In such circumstances humans are de facto out of the loop" (Heyns 2013: para. 41). That scenario would still leave the commander with the duty to deactivate or override an autonomous drone as soon as practicable after becoming aware of its misuse.

This chapter has focused on the existence of an 'accountability gap' with respect to violations of IHL as a result of the use of drone technology. In her thoughtful chapter, Chantal Meloni argues that 'reading IHL together with human rights law provisions would produce a rule that does require states to prefer the apprehension of enemies/suspected terrorists over killing them.' Perhaps this is an expression of *lex ferenda*,[76] but it is an incorrect statement of the *lex lata*. Today, during armed conflict, IHL is designed to regulate the specific conduct of hostilities, while International Human Rights Law (IHRL) continues to function when appropriate. Although, in certain situations, IHRL may still be applicable, separately or jointly with IHL (ICJ *Reports* 2004: 136), enemy combatants, whether in International Armed Conflict or in Non-international Armed Conflict, enjoy no protection from the use of lethal force, unless he/she is hors du combat or taken prisoner. Nor is there any law or rule of IHL that 'requires states to prefer' (an awkwardly worded duty) 'the apprehension of enemy combatants'[77] Professor Meloni suggests that 'we have to deal with a mixed system composed by IHL and human rights principles.' While that statement may be literally true, it is not correct to suggest that IHL and IHRL must be 'mixed' together in every situation. In a battle scenario, when soldiers are killing other soldiers (and trying not to kill civilians under the IHL principle of distinction), IHL will apply. In other circumstances, such as occupied territories, when issues arise concerning freedom of expression and/or religion and/or fair trial and/or movement of inhabitants of the occupied zone, both the IHL law of occupation and IHRL must be considered.

Similarly, Professor Meloni concludes that a 'thorough legal analysis would show that most of the targeted killings conducted outside the battlefield (and therefore regulated by the law enforcement paradigm) constitute premeditated deprivations of life that can only rarely be justified with unprovoked, absolute, and immediate necessity within the meaning of self-defence.' The key phrase in that statement is 'outside the battlefield.'

What is the scope of 'the battlefield' in the West's armed conflict against al-Qaeda? (assuming that it is an armed conflict).[78] There is at least a rational argument to be made that al-Qaeda has sufficient organizational capacity to constitute an 'organized armed group.' While this classification normally applies to Non-international Armed Conflicts, it would defeat the purpose of IHL to suggest that an armed conflict cannot exist between a state and an organized armed group located outside the state – as Meloni's discussion of the Shehadeh case would indicate. All parties seem to agree that the Shehadeh home was within the 'battlefield'; the question – under IHL, not International Human Rights Law – was whether the attack was indiscriminate. Under IHL, an enemy soldier can be attacked at any time, whether he is 'on the battlefield' or having a shower behind the front lines.[79] Under that (IHL) paradigm, many of the targeted killings conducted since 2001 in Afghanistan, Pakistan, Yemen, and elsewhere may constitute lawful attacks, as long as the legal obligations concerning distinction and precautions-in-attack have been fulfilled.

4. Conclusions

The argument that the development and employment of autonomous drones will result in an 'accountability gap' presupposes that prosecutions for violations of complex targeting rules of API – such as 'proportionality' – during armed conflict are common; they are not (Modirzadeh 2014: note 121).[80] 'Close cases' involving hard moral and legal choices by military commanders will not easily produce the culpable *mens rea* required for individual criminal responsibility. 'In determining whether an attack was proportionate, it is necessary to examine whether a reasonably well-informed person in the circumstances of the actual perpetrator, making reasonable use of the information available to him or her, could have expected excessive civilian casualties to result from the attack' (ICTY *Galić* Trial, supra note 51: para. 58).[81] Thus, the multilayered nature of many targeting decisions made during the stress of combat serves as an intellectual bar to allegations of criminal misconduct.

The discussion in this chapter has demonstrated that the development and use of autonomous drones will not result in a dramatic new 'accountability gap' for serious violations of international law. Indeed, due to the ability of advanced electronic technology to record its activities, employment of autonomous drones may provide new evidentiary avenues for determining the individual criminal responsibility of operators and commanders.

Notes

1 'Individual criminal responsibility reflects the particular degree of blameworthiness of an act committed by a moral agent' (Decoeur 2013: 480).
2 Boothby (2012: 93) explains that a commander's choice of means and method of attack will depend on many factors, 'including the nature of the objective, whether it is static or mobile, whether the civilians or civilian objects are static or mobile, what is known

about the circumstances in the area of planned attack shortly before it takes place, and whether the picture is made more complex by actions of the adverse party, perhaps including the use of human shields.'

3 The United States' Low-Cost Autonomous Attack System (LOCAAS), operates in swarms, i.e. groups of airborne 'intelligent munitions' that autonomously search for and destroy mobile targets over a wide area (U.S. Department of Defense 2013: 78).

4 The principle of individual guilt requires that an accused can be convicted of a crime only if her *mens rea* comprises the *actus reus* of the crime; a conviction absent *mens rea* would violate the presumption of innocence (ICTY *Naletilić & Martić* 2006: para. 114). Thus, to be convicted, an accused of a crime must, at a minimum, have had knowledge of the facts that made her conduct the crime. Similarly, at the ICC, conviction can occur 'only if the material elements are committed with intent and knowledge' (ICCSt 1998: Art. 30). This conjunctive approach requires the accused to possess a volitional element encompassing two possible situations: (1) she knows that her actions or omissions will bring about the objective elements of the crimes, and she undertakes such actions or omissions with the express intent to bring about the objective elements of the crime, or (2) although she does not have the intent to accomplish the objective elements of the crime, she is nonetheless aware that the consequence will occur in the ordinary course of events (ICC *Katanga & Ngudjolo* 2008: para. 529; ICC *Banda & Jerbo*: para. 153).

5 The principle of distinction 'is the foundation upon which the codification of the laws and customs of war rests' (International Committee of the Red Cross 2012: para. 1863). Thus, customary international law provides an absolute prohibition on the targeting of civilians; military necessity will not provide a basis for derogation from this prohibition (ICTY *Galić* 2006: para. 130). In addition to attacks directed against civilians, '[a] cts or threats of violence the primary purpose of which is to spread terror among the civilian population are prohibited' [Geneva Conventions Additional Protocol I 1977: Art. 51(2)]. The object and purpose of Article 51(2) is to confirm the customary rule that civilians must enjoy general protection against the danger arising from hostilities as well as the customary prohibition against attacking civilians (ICTY *Galić* 2006: para. 103).

6 The Geneva Convention (Geneva Convention API 1977: Art. 52) defines 'military objectives' as 'those objects which by their nature, location, purpose or use make an effective contribution to military action and whose total or partial destruction, capture or neutralization, in the circumstances ruling at the time, offers a definite military advantage.' In their Commentary to Art. 51, ICRC observes that military objectives principally include 'the armed forces, their members, installations, equipment and transports' (ICRC Commentary: para. 1951). Limited areas of strategic physical space, such as bridgeheads or mountain passes may, in certain circumstances, qualify as military objectives (ICRC Commentary: para. 1955).

7 Attacks that employ certain means of combat that cannot discriminate between civilians and civilian objects and military objectives are 'tantamount to direct targeting of civilians' (ICTY *Strugar* 2008: note 689, citing ICTY *Galić*: note 101). Similarly, encouragement of soldiers to fire weapons for which they lack training may be indicative of the indiscriminate nature of an attack (ICTY *Strugar* 2008: para. 274). Furthermore, the indiscriminate nature of an attack may be circumstantial evidence that the attack actually was directed against the civilian population (ICTY *Galić*: para. 132).

8 Importantly, a belligerent party may, in extreme cases of military necessity, destroy objects that *are* indispensable to the survival of the civilian population in portions of its territory that are under its control (ICRC Commentary: para. 2121). Article 54 of Geneva Convention Additional Protocol I was drafted before the development of nationwide and global computer networks that operate and maintain vital communication, transportation, electrical, and defence systems. Whether these networks should be considered as objects indispensable to the survival of the civilian population will be considered in the section on individual criminal responsibility and autonomous cyber weapons.

9 As technology develops, the scope of what is 'practicable,' and therefore legally necessary, may expand accordingly (Beard 2009: 433–439).

10 Electronic mail message from Prof. Dr. Michael Bothe, member of the German delegation to the Diplomatic Conference on the Reaffirmation and Development of International Humanitarian Law [(1974–1977), 22 May 2014, author's possession]. See ICRC (ICRC Commentary: para. 2197). Thus, upon its ratification of API, the UK government observed that '[m]ilitary commanders and others responsible for planning, deciding upon or executing attacks necessarily have to reach decisions on the basis of their assessment of the information from all sources which is reasonably available to them at the relevant time.' (Boothby 2012: note 3 at 89, nte. 59, quoting statement made by the UK on ratification of API on 28 January 1998).

11 Indeed, the ICRC Commentary to Article 57 observes that '[w]hat is required of *the person* launching an offensive is to take the necessary identification measures in good time in order to spare the population as much as possible' (ICRC Commentary: para. 2198, emphasis added).

12 According to UK military doctrine, when considering the means or methods of attack to be used, 'a commander should have regard to the following factors: (a) the importance of the target and the urgency of the situation; (b) intelligence about the proposed target – what it is being, or will be, used for and when; (c) the characteristics of the target itself, for example, whether it houses dangerous forces; (d) what weapons are available, their range, accuracy, and radius of effect; (e) conditions affecting the accuracy of targeting, such as terrain, weather, and time of day; (f) factors affecting incidental loss or damage, such as the proximity of civilians or civilian objects in the vicinity of the target or other protected objects or zones and whether they are inhabited, or the possible release of hazardous substances as a result of the attack; (g) the risks to his own troops of the various options open to him' (U.K. Joint Doctrine and Concepts Centre 2004: para. 5.32.4).

13 'Concrete and direct' means that the advantage to be gained is identifiable and quantifiable and one that flows directly from the attack, not some pious hope that it might improve the military situation in the long term' (U.K. Joint Doctrine and Concepts Centre 2004: para. 5.33.3). The Rome Statute for the International Criminal Court creates a higher standard for culpability for violations of the proportionality rule. According to Art. 8(2)(b)(iv), such attacks must be '*clearly* excessive in relation to the concrete and direct overall military advantage anticipated' (emphasis added). This language appears to contradict the ICRC's interpretation of the proportionality rule, which would prohibit 'extensive' civilian harm, when 'clearly excessive' or not: 'the Protocol does not provide any justification for attacks which cause extensive civilian losses and damages. Incidental losses and damages should never be extensive' (ICRC Commentary: para. 1980).

14 Questions that may impact a commander's proportionality analysis include, 'What are the relative values to be assigned to the anticipated military advantage gained and the expected injury to non-combatants and/or damage to civilian objects? What do you include or exclude in calculating these values? To what extent is a military commander obligated to expose her own forces to danger in order to limit civilian casualties or damage to civilian objects?' (ICTY Final Report: para. 49). Since different commanders possess different doctrinal backgrounds and different levels of combat experience, this report suggested that the standard to apply for assessing past proportionality determinations should be that of the 'reasonable military commander' (ibid.: para. 50).

15 In *Galić*, the Trial Chamber held that, when determining whether an attack was proportionate, 'it is necessary to examine whether a reasonably well-informed person in the circumstances of the actual perpetrator, making reasonable use of the information available to him or her, could have expected excessive civilian casualties to result from the attack' (ICTY *Galić*: paras. 58–60).

16 Such localities must 'fulfill the following conditions: (a) all combatants, as well as mobile weapons and mobile military equipment, must have been evacuated; (b) no hostile use shall be made of fixed military installations or establishments; (c) no acts of hostility shall be committed by the authorities or by the population; and (d) no activities in support of military operations shall be undertaken' [Geneva Convention API: Art. 59 (2); Convention (IV) (1907: Art. 25); ICCSt: Art. 8(2)(b)(v)].

17 For example, the rules of API requiring soldiers to distinguish between combatants and civilians when planning and executing attacks, to refrain from indiscriminate and disproportionate attacks, and to exercise precautionary measures prior to and during attacks have become part of customary international law (ICRC Customary IHL: Rules 1–24).

18 Cf ICJSt [Art. 38(1)(b)]. The ICJ has held that:
[n]ot only must the acts concerned amount to a settled practice, but they must also be such, or be carried out in such a way, as to be evidence of a belief that this practice is rendered obligatory by the existence of a rule of law requiring it. The need for such a belief, i.e., the existence of a subjective element, is implicit in the very notion of the *opinio juris sive necessitatis*. The States concerned must therefore feel that they are conforming to what amounts to a legal obligation. The frequency, or even habitual character of the acts is not in itself enough. (ICJ *North Sea Continental Shelf Cases*: para. 77).

19 See also European Convention for the Protection of Human Rights and Fundamental Freedoms (Art. 7), stating that no one shall be convicted of a crime on account of any act or omission that did not constitute a criminal offence under national or international law at the time when it was committed.

20 Cf. Geneva Convention API (Arts. 85–87) concerning 'Grave Breaches' of the provisions of API. The grave breaches provisions of API are based on the universality principle of jurisdiction. This principle requires states either to prosecute before their own courts or to extradite persons charged with responsibility for grave breaches [Meron 1989: 209; ICCSt: Art. 8(2)(b) and (e)].

21 As William Fenrick observes, '[C]ombat involves both lawful and unlawful killing, injury and destruction. . . . Unlawful attacks are those which are: (i) directed against civilian objects; (ii) indiscriminate; or (iii) directed against military objectives in circumstances where it can be reasonably anticipated that excessive (disproportionate) death or injury will be inflicted on civilians and/or excessive (disproportionate) damage will be caused to civilian objects' (Fenrick 2004: 157).

22 In the 'Čelebići' case, the ICTY Appeals Chamber held that the superior 'would be tried for failure to act in respect of the offences of his subordinates *in the perpetration of which he did not directly participate*' (ICTY *Zejnil Delalić* et al.: para. 225, emphasis added).

23 See, for example, ICCSt (Art. 7) concerning crimes against humanity.

24 See ICC *Bemba* (para. 427): decision pursuant to Article 61(7)(a) and (b) of the Rome Statute on the Charges of the Prosecutor Against Jean-Pierre Bemba Gombo; see also ITCY *Delalić et al.* (para. 425).

25 The Rome Statute, however, excludes the application of the *dolus eventualis* standard, as well as the *mens rea* of recklessness, at the ICC [Judgment, *Thomas Lubanga Dyilo* (ICC-01/04–01/06, 14 March 2012, para. 1011)]. At the ICC, the criminal *mens rea* exists if the accused means to commit the crime, or if he is aware that by his actions or omissions, the crime will occur in the ordinary course of events (Ibid.: para. 1018).

26 ICCSt, Art. 25(3)(d); Lubanga, supra note 57, para. 1018.

27 Ibid. Article 25 of the Rome Statute does not recognize 'planning' as a mode of liability. The drafters of Article 25 wished to include modes of liability accepted by most major legal traditions. 'Planning' as a mode of liability does not commonly appear in the continental legal system' (Email communication with Dr. Fabricio Guariglia,

Senior Appeals Counsel, Office of the Prosecutor, ICC, 29 January 2012). However, most conduct recognized as 'planning' at the ad hoc tribunals would incur liability under the broad categories of responsibility described in Article 25(3)(c) and (d). (Boas et al. 2007: 371).

28 ICRC Commentary to Article 87, API, para. 3550.

29 A superior's actual knowledge that his subordinates were committing or were about to commit a crime may be established by circumstantial evidence, including the number, type, and scope of illegal acts; time during which the illegal acts occurred; number and types of troops and logistics involved; geographical location; whether the occurrence of the acts is widespread; tactical tempo of operations; modus operandi of similar illegal acts; officers and staff involved; and location of the commander at the time (Ibid., para. 524. See also *Bemba*, supra note 56, para. 431).

30 It is not required that the superior actually acquainted herself with the information (ICTY Delalić: 239; ICC Bemba: para. 432–433).

31 Factors that may be relevant to the assessment of a superior's discharge of her duty to prevent crimes include (1) failure to secure reports that military actions have been carried out in accordance with international law, (2) failure to issue orders intended to bring the relevant practices into accord with the rules of war, (3) failure to protest against or to criticize criminal action, (3) the failure to take disciplinary measures to prevent the commission of atrocities by the troops under the superior's command, and (4) the failure to insist before a superior authority that immediate action be taken (ICTY Limaj: para. 528). With respect to the second obligation imposed by the doctrine of superior responsibility, a superior's duty to punish perpetrators of crime encompasses the obligation to conduct an effective investigation with a view to establishing the facts (Ibid.: para. 529; ICC Bemba: para. 438). The superior's obligation is to take active steps to ensure that perpetrators will be punished (ICC Limaj: para 529). 'To that end, the superior may exercise his own powers of sanction, or if he lacks such powers, report the perpetrators to the competent authorities' (Ibid.).

32 At the International Criminal Court, proof of the existence of the superior's effective control over subordinates who committed crimes requires temporal evidence demonstrating that the superior's effective control existed at least when the crimes were about to be committed (ICC Bemba: para. 419).

33 This standard requires more of an active duty on the part of the commander to take the necessary measures to secure knowledge of the conduct of her troops and to inquire, regardless of the availability of information at the time on the commission of the crime (ICC Bemba: para. 433).

34 ICCSt Art. 28(1)(a).

35 ITCY Bemba, para. 432.

36 ICCSt Art. 28(2)(a) (emphasis added).

37 Bemba, supra note 56, para. 439.

38 Ibid.

39 ICCSt Art. 28(1)(b) and 28(2)(c). Assessments of measures taken by superiors do not depend on whether they 'were of a disciplinary or criminal nature' but rather whether they were necessary or reasonable in the circumstances of the case (Bemba, supra note 56, para. 495).

40 ICTY *Prosecutor v. Stanislav Galić* (IT-98–29-T), 5 December 2003, paras. 251, 256, 269, 274, 287, 316, 320, 336, 344, and 405.

41 Ibid., paras. 57, 569–577; *Galić* Appeals Judgment, supra note 15 at paras.131–134.

42 On 21 February 2010, nearly 30 Afghan civilians were killed or wounded after ISAF helicopters fired on a convoy of vehicles suspected of containing Taliban soldiers approaching U.S. forces. A Predator drone, flown and monitored by a crew in the U.S., surveilled the vehicles for hours prior to the airstrike. The Predator ground crew failed to accurately inform the ground commander of the possible presence of civilians in

the vehicles. An after-action inquiry into the incident concluded that the ground commander 'displayed tactical patience in letting the situation develop for several hours before the engagement. The time bought by that patience was wasted because of the Predator crew's inaccurate reporting' and the failure of pertinent commanders to properly analyze the situation and provide insights, analysis, or options to the ground commander (USFOR-ADCDR-S 2010: para. 5).

43 In modern warfare, the term 'tactical level' refers to the 'level at which activities, battles and engagements are planned and executed to accomplish military objectives assigned to tactical formations and units.' Thus, a 'tactical commander' exercises the authority 'to assign tasks to forces under his command for the accomplishment of the mission assigned by higher authority' (NATO 2014: 2-T-1–2-T-2).

44 M. Schmitt, remarks during panel on The International Legal Context at Autonomous Military Technologies: Policy and Governance for Next Generation Defence Systems, Chatham House, London, 24 February 2014; permission to cite in electronic mail message to author, 15 March 2014.

45 See Corn (2014: 6–7 and 9–11) concerning the duties of the 'compliance commander' vis-à-vis autonomous weapon systems.

46 Cf. Müller (this volume).

47 Chambers at the ICTY and the ICC have held that '(. . .) [t]he more physically distant the superior was from the commission of the crimes, the more additional indicia are necessary to prove that he knew them' (ICC Bemba: para. 484, citing ICTY Stakić: para. 460). In Bemba, the evidence established that Jean-Pierre Bemba, the president of the Mouvement de Libération du Congo (MLC), in his effective role as commander of MLC troops, communicated directly with his commanders in the field, had access to a regular reporting system, and regularly monitored international media reports about the events in the Central African Republic (ICC Bemba: paras. 486 and 488–489).

48 For example, in an iconic photograph taken in the White House Situation Room in Washington, D.C., on 1 May 2011, U.S. President Obama and members of his staff are seen intently watching 'live updates' of Operation Neptune Spear, which resulted in the death of Osama Bin Laden in Pakistan (Situation Room, available online at http://en.wikipedia.org/wiki/Situation_Room). Moreover, the use of Youtube by all sides of the Syria conflict provides a graphic video record of events in Syria within minutes of their occurrence. https://www.youtube.com/results?search_query=syria+bread+line+air+strike; https://www.youtube.com/results?search_query=syria+rebels+linked+to+al+qaeda+apologize+for+beheading+fellow+fighter.

49 See United Nations War Crimes Commission (1947: 93f) (The Zyklon B Case). Temporal distance in another sense – the time between the alleged crimes and the commencement of prosecution – will not bar prosecutions of persons responsible for unlawful attacks employing autonomous weapon systems, as the imprescriptibility of war crimes, crimes against humanity, and genocide is part of customary law (Simma and Paulus 1999: 315).

50 I am grateful to Professor Geoffrey Corn for his insights on this topic.

51 The control algorithms that comprise artificial intelligence for autonomous systems are created and tested by teams of humans (U.S. Department of Defense 2013: 67). Such high testing and validation standards would detect negligent or intentional 'mis-programming' of component systems of autonomous drones by computer programmers and designers of these weapon systems. Detection and correction of such mistakes before the autonomous drone reaches the battlespace would vitiate concerns about how to hold scientists, engineers, computer programmers, and others accountable when the autonomous drones that they help to design and manufacture subsequently are employed in violations of law.

52 One definition of 'strict criminal liability' entails offences where, with respect to at least one element of the *actus reus*, 'the offender's mental state is irrelevant' (Prendergast

2011: 286). Some commentators argue that strict criminal liability violates international human rights norms, such as the rights to a fair trial, in particular the presumption of innocence (Salako 2006: 537, 549). Furthermore, the ICTY Appeals Chamber implicitly rejected the use of strict criminal liability when it observed that the principle of individual criminal responsibility 'requires that fundamental characteristics of a war crime be mirrored in the perpetrator's mind' (ICTY Naletilić & Martinović: para. 118).

53 Well-documented *absences* of accountability and/or criminal responsibility are not uncommon in modern warfare. For example, in September 2009, Colonel Georg Klein, a German commander serving in ISAF in Kunduz province in Afghanistan, ordered an airstrike on a fuel tanker that had been stolen by members of the Afghan Taliban. The strike allegedly killed over a hundred civilians. In April 2010, German prosecutors, citing Colonel Klein's lack of knowledge of the presence of civilians at the bombing site, declined to charge Colonel Klein for criminal responsibility for the civilian deaths (Houlton 2010). In addition, during the 1999 NATO bombing campaign, hundreds of civilians died in multiple incidents including aerial attacks on a civilian passenger train, a convoy of civilians, a radio and television station, among others. A subsequent report by the ICTY Office of the Prosecutor concluded that 'neither an in-depth investigation related to the bombing campaign as a whole nor investigations related to specific incidents are justified. In all cases, either the law is not sufficiently clear or investigations are unlikely to result in acquisition of sufficient evidence to substantiate charges against high level accused or against lower accused for particularly heinous offences' (ICTY Final Report 2000).

54 Cf. Müller (this volume), arguing that the presence of responsibility gaps does not constitute in itself an argument against the use autonomous drones.

55 In December 2013, a German District Court rejected a civil lawsuit for damages brought against the German government by the families of victims of the airstrike at Kunduz (Spiegel Online International 2013).

56 In 2003, for example, one highly respected jurist expressed his concern that 'the gap between the achievable and the hoped-for' in international criminal justice could undermine the effectiveness of institutions such as the ICC (Fulford 2003).

57 Customary International Humanitarian Law provides that '[i]ndividuals are criminally responsible for the war crimes they commit' (ICRC Customary IHL: Rule 151).

58 During the Clinton administration, after U.S. armed forces under his command launched cruise missiles against the headquarters of Saddam Hussein's intelligence service in Baghdad, President Clinton was aghast to learn that the missiles neither had cameras mounted on them, nor could they be 'turned back' prior to striking their targets (Clarke 2004: 82–83).

59 'In considering whether commanders and others responsible for planning, deciding upon, or executing attacks have fulfilled their responsibilities, it must be borne in mind that they have to make their decisions on the basis of their assessment of the information from all sources which is available to them at the relevant time' (UK Joint Doctrine and Concepts Centre 2004: 85). This means looking at the situation as it appeared to the individual at the time when she made her decision [Ibid. Cf. Schmitt (2013: note 384)].

60 Professor Schmitt concisely describes the potential legal consequences for commanders who deploy autonomous weapon systems without full comprehension of their technology and capacities: 'you can't use a weapon in the battlespace that you don't understand; that's an indiscriminate use of the weapon. That's a war crime' (M. Schmitt, remarks during panel on The International Legal Context at Autonomous Military Technologies: Policy and Governance for Next Generation Defence Systems, Chatham House, London, 24 February 2014; email message, 15 March 2014). Similarly, as already mentioned, encouragement of soldiers to fire weapons for which they lack training may be indicative of the indiscriminate nature of an attack (ICTY Strugar: para. 274; Boothby 2012: 92).

61 As discussed, the Rome Statute excludes the application of the *dolus eventualis* standard, as well as the *mens rea* of recklessness, at the ICC (ICC *Thomas Lubanga Dyilo*: para. 1011). The ad hoc tribunals, however, will convict accused under this standard (ICTY *Tihomir Blaškić*: paras. 39–42).

62 As the autonomy of weapons systems increases, the onus on a commander will be to know her machines as much as the men and women who serve as her subordinates. Subordinates will be, for the most part, technicians and engineers rather than armed combatants. This evolution will naturally lead to changes in the climate and culture of command in modern armed forces. NATO doctrine encourages commanders to create a climate that encourages subordinates to think independently and to take initiative (NATO 2010: paras. 6–14). It remains an open question whether introduction of autonomous weapons systems will reduce independent thinking and initiative on the part of human soldiers.

63 B. Burridge, 'Military Capability Is Founded on a Body of Knowledge to Which Industry Is a Major Contributor,' 2010, 2, copy in author's possession. 'In counter-IED and in some weapon applications, it is not unusual for commanders to get advice directly from the [manufacturer] concerned and vice versa. In many cases, this results in a software tweak hence the virtual communication ultimately runs back to the software engineer but inside the business. Contrast this with buying off-the-shelf from a company who might not regard your query as either important or relevant' [Air Chief Marshal (ret.) Sir Brian Burridge, former commander of U.K. forces in Iraq, email communication with author, 17 March 2014].

64 Furthermore, in November 2012, the ICTY Appeals Chamber overturned the conviction of Croatian General Ante Gotovina, partly on the basis of the Trial Chamber's unsubstantiated finding that an artillery attack on Milan Martić was disproportionate. The majority decision noted that the Trial Chamber's proportionality analysis 'was not based on a concrete assessment of comparative military advantage, and did not make any findings on resulting damages or casualties' (ICTY Gotovina: para. 82).

65 According to the U.S. Department of Defense, '[t]he current state of autonomy for most tactical unmanned ground systems requires human decision makers and line-of-sight communications. Systems that are autonomous require highly structured and predictable environments' (U.S. Department of Defense 2013).

66 In contrast to fully preprogrammed systems, autonomous weapon systems that utilize machine learning will develop modified strategies for themselves by which they select their behavior (Ibid.: 67).

67 Recent modifications of the LOCAAS swarm concept introduced greater human supervision to 'enable retargeting and the ability to abort attack by a human controller when required' (Ibid.: 78).

68 States would be unlikely to make the massive economic investments required to develop and deploy such weapons without this expectation.

69 The ICRC takes a similar view: '. . . it is the duty of parties to the conflict to have the means available to respect the rules of the Protocol. . . . [I]t is reprehensible for a Party possessing such means not to use them, and thus consciously prevent itself from making the required distinction' (ICRC Commentary to API: para. 1952).

70 This premise should hold true regardless of whether courts apply the had-reason-to-know standard of the ad hoc tribunals or the should-have-known standard of the ICC.

71 For example, during multinational NATO operations, the Supreme Allied Commander for Europe (SACEUR) must '[e]stablish an intelligence architecture linking NATO Headquarters with national intelligence centres to provide the [Joint Force Commander] with a common, timely and accurate picture of the situation during all phases of the campaign' [NATO 2010: 0615(e)].

72 ICRC Commentary to Art. 87, API, para. 3560. Available online at http://www.icrc.org/applic/ihl/ihl.nsf/Comment.xsp?viewComments=LookUpCOMART&articleUNID=C203555D5F233D81C12563CD0051E1FE.

73 Lockheed Martin, 'VCS-4586 Capabilities Guide: Ground Control Operator Software for Unmanned Vehicle Systems.' Available online at http://www.lockheedmartin.com/content/dam/lockheed/data/ms2/documents/cdl-systems/VCS-4586%20CAPABILITIES%20GUIDE-August2013.pdf.
74 Ibid, para. 527; ICC *Bemba*, para. 437.
75 One notable exception to this attribute may be certain autonomous cyber weapons that, once released into the Internet, may lack a specific deactivation mechanism or are encrypted to prevent outside deactivation (Clayton 2012b). Contrast this characteristic with the infamous Stuxnet malware, which was programmed to cease its operations in late June 2012 (Clayton 2012a). Or such weapons may be incorporated into computer hardware at the time of manufacture, creating the capacity to disrupt or deactivate the computer system (Markoff 2009).
76 The term '*lex ferenda*' is an expression used to refer to what the law should be, as opposed to *lex lata*, which refers to the law as it currently exists.
77 The ICRC has promoted a similar 'guideline' that does not serve as a statement of the law (ICRC Interpretive Guidance). For criticism of the ICRC Guidance, see Parks (2010); Watkin, (2010); Schmitt (2010a). For a more positive view, see the remarks of Goodman (2010).
78 For a discussion of the components of international and non-international armed conflicts, see ICTY Tadić (para. 70); ICTY Limaj (paras. 83–173); ITCY *Boškoski & Tarčulovski* (paras. 173–292).
79 In April 1943, for example, U.S. intelligence intercepted the flight plan of Admiral Isoroku Yamamoto, Commander of the Japanese Navy, for a visit to the Solomon Islands. U.S. Navy planes shot down Yamamoto's plane above the South Pacific. This author is unaware of suggestions that the killing of Admiral Yamamoto occurred 'outside the battlespace' or within a 'law enforcement paradigm' (Liddell Hart 1970: 503).
80 For example, at the ICC, '[t]he Statute definition of the offense of launching a disproportionate attack adds the word "clearly" to the text of the provision as it appears in Additional Protocol I. This addition is intended to indicate to the Court that only obvious cases of disproportionate attacks should be punished' (Newton and May 2014: 151, citing Gardham 2004: 134). Thus, the ICC Statute leaves a broad margin of discretion to belligerents (Newton and May 2014: note 144, 162).
81 Marco Sassóli (2012: 335) observes that the *Galić* judges attempted to create a more objective test for evaluation of proportionality decisions.

References

Beard, J. 2009. Law and War in the Virtual Era. *American Journal of International Law* 103(3): 409–445.

Boas, G, JL Bischoff and NL Reid. 2007. *Forms of Responsibility in International Criminal Law. International Criminal Law Practitioners Library*, Vol. I. Cambridge University Press.

Boothby, W. 2012. *The Law of Targeting*. Oxford University Press.

Burridge, B. 2009. *U.K. Inquiry into Iraq ('Chilcot Inquiry')*, 8 December. Available online at http://www.iraqinquiry.org.uk/media/39393/091208burridge-brims.pdf.

Clarke, R. 2004. *Against All Enemies: Inside America's War on Terror*. Free Press.

Clayton, M. 2012a. Stuxnet Cyberweapon Set to Stop Operating. *The Christian Science Monitor*, 23 June. Available online at http://www.csmonitor.com/USA/2012/0623/Stuxnet-cyberweapon-set-to-stop-operating.

———. 2012b. More Telltale Signs of Cyber Spying and Cyber Attacks Arise in Middle East. *The Christian Science Monitor*, 21 August. Available online at http://www.csmonitor.com/USA/2012/0821/More-telltale-signs-of-cyber-spying-and-cyber-attacks-arise-in-Middle-East-video.

Convention (IV). 1907. *Convention (IV) Respecting the Laws and Customs of War on Land and Its Annex: Regulations Concerning the Laws and Customs of War on Land.* The Hague, 18 October. Available online at http://www.icrc.org/applic/ihl/ihl.nsf/Article.xsp?action=openDocument&documentId=D1C251B17210CE8DC12563CD0051678F.

Corn, G. 2014. Panel discussion, Article 36 Legal Review and Legal Compliance, Conference on Autonomous Weapons Systems – Law, Ethics, Policy. European University Institute, Academy of International Law, Florence, 25 April; associated draft paper, 'Autonomous Weapon Systems: Managing the Inevitability of "Taking the Man Out of the Loop."' Copy in author's possession.

Cummings, M. 2006. Automation and Accountability in Decision Support System Interface Design. MIT Human and Automation Laboratory. Available online at http://web.mit.edu/aeroastro/labs/halab/papers/Cummings_JTS.pdf.

———. 2014. Man versus Machine or Man + Machine? *Intelligent Systems, IEEE* 29(5): 62–69.

Decoeur, H. 2013. Avoiding Strict Liability in Mixed Conflicts: A Subjectivist Approach to the Contextual Element of War Crimes. *International Criminal Law Review* 13(2): 473–492.

Dinstein, Y. 2010. *The Conduct of Hostilities under the Law of International Armed Conflict, 2nd ed.* Cambridge University Press.

European Convention for the Protection of Human Rights and Fundamental Freedoms. (n.d.). *European Convention on Human Rights.* See pp. 31–33. Available online at http://www.echr.coe.int/Documents/Convention_ENG.pdf.

Fenrick, WJ. 2004. The Prosecution of Unlawful Attack Cases before the ICTY. *Yearbook of International Humanitarian Law* 7: 153–189.

Fiddian, P. 2012. UAV Swarm Technology Trial Success. *Armed Forces International News* 7 August. Available online at http://www.armedforces-int.com/news/uav-swarm-technology-trial-success.html.

Fulford, A. 2003. Foreword, in I Bantekas & S Nash (eds), *International Criminal Law*, 2nd ed. Cavendish Publishing.

Gardham, J. 2004. *Necessity, Proportionality, and the Use of Force by States.* Cambridge University Press.

Geneva Conventions API. 1977. *Protocol Additional to the Geneva Conventions of 12 August 1949, and Relating to the Protection of Victims of International Armed Conflicts (Protocol I)*, 8 June. Available online at https://www.icrc.org/applic/ihl/ihl.nsf/Treaty.xsp?documentId=D9E6B6264D7723C3C12563CD002D6CE4&action=openDocument.

Garraway, C. 2010. *The Changing Character of the Participants in War: Civilianization of War-Fighting and the Concept of "Direct Participation in Hostilities."* U.S. Naval War College, International Law Conference. Available online at https://www.usnwc.edu/Research---Gaming/International-Law/New-International-Law-Studies-%28Blue-Book%29-Series/International-Law-Blue-Book-Articles.aspx?Volume=87.

Henckaerts, JM, and Doswald-Beck, L. (eds). 2005. *Customary International Humanitarian Law.* Cambridge University Press, at pp. 25–71 and 551–563.

Heyns, C. 2013. *Report of the Special Rapporteur on Extra-Judicial, Summary or Arbitrary Executions*, A/HRC/24/47, 9 April, paras. 89–97.

Houlton, S. 2010. German Prosecutors Drop Case against Kunduz Airstrike Colonel. *DW* 19 April. Available online at http://www.dw.com/en/german-military-drops-case-against-kunduz-airstrike-colonel/a-5926249.

ICC *Banda & Jerbo*. Corrigendum to Decision on Confirmation of Charges (ICC-02/05–03/09), Pre-Trial Chamber I, 7 March 2011.

ICC *Bemba*. Decision Pursuant to Article 61(7)(a) and (b) of the Rome Statute on the Charges of the Prosecutor against Jean-Pierre Bemba Gombo, *Bemba* (ICC-01/05–01/08), Pre-Trial Chamber II, 15 June 2009.

ICC *Katanga & Ngudjolo*. Decision on Confirmation of Charges, Pre-trial Chamber. ICC-01/04–01/07, 30 September 2008.

ICC *Thomas Lubanga Dyilo*. Judgment. ICC-01/04–01/06, 14 March 2012.

ICCSt. 1998. *Rome Statute of the International Criminal Court*. A/Conf.183/9, 17 July 1998. Available online at http://www.icc-cpi.int/en_menus/icc/legal%20texts%20and%20tools/official%20journal/Pages/rome%20statute.aspx.

ICJ *North Sea Continental Shelf Cases*. Judgment, ICJ Reports, 20 February 1969.

ICJ *Reports*. 2004. Legal Consequences of the Construction of a Wall in the Occupied Palestinian Territory, Advisory Opinion. Available online at http://www.icj-cij.org/docket/files/131/1671.pdf.

ICJSt. N.d. *Statute of the International Court of Justice*. Available online at http://www.icj-cij.org/documents/?p1=4&p2=2.

ICRC *Commentary API*. International Committee of the Red Cross, 'Protocol Additional to the Geneva Conventions of 12 August 1949, and Relating to the Protection of Victims of International Armed Conflicts ("API"), 8 June 1977.' Accessed at para. 1863. Available online at https://www.icrc.org/applic/ihl/ihl.nsf/Treaty.xsp?documentId=D9E6B6264D7723C3C12563CD002D6CE4&action=openDocument.

ICRC *Commentary References*. Available online at https://www.icrc.org/applic/ihl/ihl.nsf/Comment.xsp?action=openDocument&documentId=78C0DA9A7B459ACEC12563CD0042F649.

ICRC *Commentary to Art. 57*. API, para. 2197. Available online at http://www.icrc.org/applic/ihl/ihl.nsf/Comment.xsp?viewComments=LookUpCOMART&articleUNID=50FB5579FB098FAAC12563CD0051DD7C.

ICRC *Commentary to Art. 67*. Geneva Convention IV. Available online at http://www.icrc.org/applic/ihl/ihl.nsf/Comment.xsp?viewComments=LookUpCOMART&articleUNID=44E67A17D850B35CC12563CD0051BF2C.

ICRC *Customary IHL*. Available online at http://www.icrc.org/customary-ihl/eng/docs/v1_rul.

ICRC *Interpretive Guidance*. Interpretive Guidance on the Notion of Direct Participation in Hostilities under IHL. Available online at https://www.icrc.org/eng/assets/files/other/icrc-002-0990.pdf.

ICTR *Nahimana*. Judgment Appeals Chamber (ICTR-99-52-A), 28 November 2007.

ICTY *Boškoski & Tarčulovski*. Appeals Chamber (IT-04-02-A), 19 May 2010.

ICTY *Brdanin*. (IT-99-37-A), Judgment 3 April 2007.

ICTY *Delalić et al*. (IT-96-21-T), Judgment, 16 November 1998.

ICTY *Final Report*. 2000. Final Report to the Prosecutor by the Committee Established to Review the NATO Bombing Campaign Against the Federal Republic of Yugoslavia (2000). Accessed at para. 48. Available online at http://www.icty.org/sid/10052.

ICTY *Galić*. Appeal (IT-98-29-A), Appeals Chamber, 30 November 2006.

ICTY *Galić*. Trial (IT-98-29-T), Judgement 5 December 2003.

ICTY *Gotovina, et al*. (IT-06-90-A), Judgment, 16 November 2012.

ICTY *Halilović*. Judgment, (IT-01-48-T), 16 November 2005, para. 63.

ICTY *Krajišnik*. (IT-00-39-A), Judgment 17 March 2009.

ICTY *Kvočka, et al*. (IT-98-30/1-A), Judgment, 28 February 2007.

ICTY *Limaj*. (IT-03-66-T), Judgment Trial Chamber, 30 November 2005.
ICTY *Naletilić & Martinović*. 2006. (IT-98-34-A), Judgment 3 May 2006.
ICTY *Perisić*. (IT-04-81-A), Appeals Chamber, 28 February 2013.
ICTY *Popović, et al*. (IT-85-88-T), Judgment 10 June 2010.
ICTY *Stakić*. (IT-97-24-T), Judgment 31 July 2003.
ICTY *Strugar*. (IT-01-42-A), Appeals Chamber, 17 July 2008.
ICTY *Tadić*. (IT-94-1-AR72), 2 October 1995.
ICTY *Tihomir Blaškić*. Judgment, (IT-95-14/A), Appeal Decision on the Defence Motion for Jurisdiction, 29 July 2004.
ICTY *Zejnil Delalić, et al*. (IT-96-21-A), Judgment 20 February 2001.
ICTY *Zlatko Aleksovski*. (IT-95-14/1- A), Judgment 24 March 2000.
Jackson, R. 2014. Remarks in Panel on Autonomous Weaponry and Armed Conflict, Annual Meeting of American Society of International Law, 10 April. Email message to author 7 May 2014. Video available online at https://www.youtube.com/watch?v=duq3DtFJtWg&list=PLYp0ZUypbrnevQlBfMUSDG0IanrvJ3J6z&index=4.
Kunduz Bombing: German Court Drops Case over Civilian Deaths. 2013. *Spiegel Online International*, 11 December. Available online at http://www.spiegel.de/international/germany/court-says-germany-not-responsible-for-damages-in-afghanistan-attack-a-938490.html.
Liddell Hart, BH. 1970. *History of the Second World War*. G. P. Putnam's Sons.
Markoff, J. 2009. Old Trick Threatens New Weapons. *The New York Times*, 26 October. Available online at http://www.nytimes.com/2009/10/27/science/27trojan.html?pagewanted=all&_r=0.
Maurer, P. 2014. War, Protection and the Law, 2014 FCO International Law Lecture, 19 May, London. Available online at https://www.icrc.org/eng/resources/documents/statement/05-23-war-protection-and-the-law-icrc-approach-fco-2nd-international-law-lecture-19-may-2014.htm.
Meron, T. 1989. *Human Rights and Humanitarian Norms as Customary Law*. Clarendon Press.
Modirzadeh, N. 2014. Remarks in panel on Autonomous Weaponry and Armed Conflict, Annual Meeting of American Society of International Law, 10 April. Video available online at https://www.youtube.com/watch?v=duq3DtFJtWg&list=PLYp0ZUypbrnevQlBfMUSDG0IanrvJ3J6z&index=4.
NATO. 2010. *AJP-01D Allied Joint Doctrine*, North Atlantic Treaty Organization, 21 December. Available online at https://www.gov.uk/government/uploads/system/uploads/attachment_data/file/33694/AJP01D.pdf.
———. 2014. *Glossary of Terms and Definitions*, AAP-06(2014), 2-T-1–2-T-2. Available online at http://nso.nato.int/nso/ZPUBLIC/_BRANCHINFO/TERMINOLOGY_PUBLIC/NON-CLASSIFIED%20NATO%20GLOSSARIES/AAP-6.PDF.
Newton, M, and May, L. 2014. *Proportionality in International Law*. Oxford University Press.
Oeter, S. 2008. Methods of Combat, in D Fleck (ed.), *The Handbook of International Humanitarian Law*, 2nd ed. Oxford University Press, 166–230.
Parks, WH. 2010. Part IX of the ICRC "Direct Participation in Hostilities" Study: No Mandate, No Expertise, and Legally Incorrect. *International Law and Politics* 42: 770–830.
Prendergast, D. 2011. The Constitutionality of Strict Liability in Criminal Law. *Dublin University Law Journal* 33: 285–318.
Salako, S. 2006. Strict Criminal Liability: A Violation of the Convention? *The Journal of Criminal Law* 70(6): 531–549.

Sassóli, M. 2012. Autonomous Weapons and International Humanitarian Law: Advantages, Open Technical Questions and Legal Issues to Be Clarified. *International Law Studies* 90: 308–340.

Saxon, D. 2014. A Human Touch: Autonomous Weapons, DOD Directive 3000.09 and the Meaning of "Appropriate Levels of Human Judgment over the Use of Force". *Georgetown Journal of International Affairs* (Summer/Fall 2014) 15 (2): 100–109.

Schmitt, MN. 2010. The Interpretive Guidance on the Notion of Direct Participation in Hostilities: A Critical Analysis. *Harvard National Security J* 1:5–44.

Schmitt, MN. (ed.). 2013. *Tallin Manual on the International Law Applicable to Cyber Warfare*. Cambridge University Press.

Simma, B, and Paulus, AL. 1999. The Responsibility of Individuals for Human Rights Abuses in Internal Conflicts: A Positivist View. *American Journal of International Law (AJIL)* 93(2): 302–316.

Supreme Court of Canada *R v. Finta*. 1994. 1 SCR 701, 760.

Trapp, K. 2013. Great Resources Mean Great Responsibility: A Framework of Analysis for Assessing Compliance with API Obligations in the Information Age, in D Saxon (ed), *International Humanitarian Law and the Changing Technology of War*. Martinus Nijhoff/Brill, 153–170.

U.K. Joint Doctrine and Concepts Centre. 2004. JSP383, *Joint Service Manual of the Law of Armed Conflict*, para. 5.32.4. Available online at https://www.gov.uk/government/uploads/system/uploads/attachment_data/file/27874/JSP3832004Edition.pdf.

UN CCW. 2004. *Protocol II to Convention on Prohibitions or Restrictions on the Use of Certain Conventional Weapons Which May Be Deemed to Be Excessively Injurious or to Have Indiscriminate Effects*. Available online at https://www.icrc.org/eng/assets/files/other/icrc_002_0811.pdf.

United Nations War Crimes Commission. 1947. *Law Reports of Trials of War Criminals*, Vol. 1. Available online at http://www.loc.gov/rr/frd/Military_Law/pdf/Law-Reports_Vol-1.pdf.

U.S. Department of Defense. 2012. Autonomy in Weapon Systems. Policy Directive 3000.09, November 21. Available online at http://www.dtic.mil/whs/directives/corres/pdf/300009p.pdf.

———. 2013. *Unmanned Systems Integrated Roadmap* FY2013–2038. Available online at http://archive.defense.gov/pubs/DOD-USRM-2013.pdf.

USFOR-ADCDR-S. 2010. *Executive Summary for AR 15–6 Investigation, 21 February 2010 CIVCAS Incident in Uruzgan Province*, para. 5. Available online at http://www.isaf.nato.int/images/stories/File/April2010-Dari/May2010Revised/Uruzgan%20investigation%20findings.pdf.

Watkin, K. 2010. Opportunity Lost: Organized Armed Groups and the ICRC "Direct Participation in Hostilities" Interpretive Guidance. *International Law and Politics* 42(3): 641.

2 State and Individual Responsibility for Targeted Killings by Drones

Chantal Meloni

1. Introduction

Notwithstanding the variety of unmanned aerial vehicles (drones) and the multitude of their possible uses in daily life and civil activities – as in surveying or even for recreational purposes – drones today are mostly associated with their military use and lethal force operations. The use of armed drones in warfare, in particular as a means to execute State-sponsored targeted killings, raises serious concerns and, as we will see, undermines fundamental principles such as the right to life and ultimately the rule of law.

Although the targeted killings practice started well before the entry into use of drones, it is a fact that since this technology was made available, the recourse to targeted killing operations escalated. The number of people killed is huge: according to the most recent data, up to 5.500 individuals have been killed by U.S. drones strikes since 2004 in Pakistan, Yemen and Somalia – countries where the United States is not at war – not counting the strikes in Afghanistan and Iraq – where, on the contrary, there is a declared state of war.[1]

The question is whether the introduction of drones in warfare operations creates new ethical and legal issues. What I will argue is that the critical aspect is not the technology as such but rather the twist of legal principles regulating lethal operations conducted by drones. This is where States are being utterly ambiguous, blurring and expanding the limits and boundaries of concepts such as "war", "battlefield", "combatant", "immediate threat", or "terrorist". Some commentators make the argument that we are standing in a grey zone, an unregulated space for killing between war and peace. Here I confute this argument, showing that the legal principles are there and actually applicable but that some States are trying to shift the boundaries of what is legally permissible, invoking exceptional circumstances. However, the danger in insisting on the existence of such a grey zone is clear: once you create a state of exception, you cannot control who will make use of it.

As we will further elaborate, two levels of responsibility are involved in the commission of unlawful targeted killings: the State level and the individual level. As to the latter one, criminal liability can come into play. Indeed, not only targeted killings are, as we will argue, mostly unlawful, thus involving the responsibility of the State, but they can also amount to war crimes and possible crimes against humanity, with the consequence that they entail the criminal responsibility of the individuals who committed them.[2]

Thus, the question is under which conditions are targeted killings legal? But, most importantly, who can determine their lawfulness, given the lack of transparency by the governments and their refusal to allow any judicial oversight on "political" and "national security" issues? The current situation is particularly worrying; it can be even questioned whether it is possible at all to ascertain individual criminal responsibility of those who planned, ordered, or conducted the kill operation within a system that lacks any transparency and is actually designed in such a way as to condone the arbitrary deprivation of lives, when committed pursuant to a State-sponsored policy and program.

2. A Major "Accountability Vacuum"

Drones are one of the most controversial topics in the worldwide public debate. Their use in warfare and more generally in military operations that imply the recourse to lethal force raises a number of issues at different levels: political, moral, military, legal, and philosophical. As anticipated, I will focus in particular on the legal regime applicable to the so-called targeted killing practice, via both manned and unmanned vehicles, and on the problematic issues related to this practice.

This issue is not only currently catalysing worldwide attention and concerns[3] but is also a crucial one to analyse the difficult interaction of the different legal regimes that regulate modern asymmetric conflicts. As noted by Philip Alston – former UN Special Rapporteur on extrajudicial, summary, or arbitrary executions – in the UN's 2010 report:

> In recent years, a few States have adopted policies that permit the use of targeted killings, . . . including in the territories of other States. Such policies are often justified as a necessary and legitimate response to "terrorism" and "asymmetric warfare", but have had the very problematic effect of blurring and expanding the boundaries of the applicable legal frameworks – human rights law, the laws of war, and the law applicable to the use of inter-state force. Even when the laws of war are clearly applicable, there has been a tendency to expand who may permissibly be a target and under what conditions. Moreover, the States concerned have often failed to specify the legal justification for their policies, to disclose the safeguards in place to ensure that targeted killings are in fact legal and accurate, or to provide accountability mechanisms for violations. Most troubling, they have refused to disclose who has been killed, for what reason, and with what collateral consequences. The result has been the displacement of clear legal standards with a vaguely defined license to kill, and the creation of a major *accountability vacuum*.[4]

It is argued that the State-sponsored method of targeted killing raises serious concerns and undermines such principles as the presumption of innocence, the right to a fair trial, the protection of the right to life, and ultimately the rule of law.[5] The purpose of this chapter is, in particular, to analyse the legal mechanisms aiming at assessing criminal responsibility for the commission of unlawful targeted killings.

But before going into that analysis, let's first have a look at the general features of the targeted killings practice, in particular when committed through drones.

3. Origins and Definition of the Targeted Killings Practice

As noted by Alston in his 2010 Report, "targeted killing" is not a legal term. Despite the frequency with which it is invoked, it is not defined under international law, nor does it fit neatly into any particular legal framework. The term came into common usage in the year 2000, after Israel made public a policy of "targeted killings" of alleged terrorists in the Occupied Palestinian Territory during the Second Intifada.[6] Since then, the term has been used in several contexts to indicate the State-sponsored practice of eliminating "enemies", in particular in the territories of other States, both in times of peace as well as during armed conflicts.

Although in most circumstances, targeted killings do violate the right to life and can also amount to war crimes, in exceptional circumstances and within a context of armed conflict, they may be legal. Therefore the term "targeted killing" in general shall not be used as synonymous with "extrajudicial execution", "summary execution", or "assassination", which are by definition always illegal. "The means and methods of killing vary, and include sniper fire, shooting at close range, missiles from helicopters, gunships, drones, the use of car bombs, and poison":[7] Whether conducted through drones or other means, the common element of targeted killings is that lethal force is intentionally and deliberately used, with a degree of premeditation, against individuals identified in advance.[8] In a targeted killing, the specific goal of the operation is to use lethal force, which distinguishes them from unintentional, accidental or reckless killings or killings made without conscious choice. It also distinguishes them from law enforcement operations, where the resort by law enforcement authorities (i.e. the police) to lethal force for self-protection against an imminent threat can be justified, but the goal to kill the suspect/criminal cannot be the sole aim of the operation from its inception.[9] In reality, as we will see, most targeted killings do not fit either the self-defence[10] or the legitimate act-of-war paradigm. They constitute premeditated deprivations of life that can only rarely be justified with the requirement for an "imminent threat" within the meaning of self-defence.[11]

The State of Israel has the longest track on targeted killings[12] and was the first country to openly acknowledge it as a policy.[13] Targeted killings are still extensively used by Israel, especially in Gaza, both in the context of major military operations (as during the 2014 so-called operation Protective Edge, where drone strikes have been massively used)[14] and outside of the outbreak of hostilities, to eliminate Palestinian suspects allegedly in a "surgical" way.[15] Beside Israel and the United States, other countries are also believed to have resorted to targeted killings, in particular the United Kingdom[16] and Russia, but there is less information available on these strikes.[17] In any case, targeted killings are not solely used by the United States and not exclusively against terrorist threats in the global "war against terror", as confirmed by the Reports of the UN Special Rapporteurs.

Nowadays, however, the term is used in particular to indicate the U.S. policy of eliminating suspected Al-Qaida and associated terrorists by combat planes or by drones. It is a fact that, since 2001 and routinely since 2009, the United States has carried out countless deliberate and premeditated killings of declared suspected terrorists overseas, the most famous of which is the targeted killing of Osama Bin Laden:[18] many other less notorious ones have been carried out before and after that.

Numbers are not clear due to the lack of transparency by governments, but thousands of people have been killed, including many hundreds of civilians, and the dead toll is rising while you are reading this. A single strike in Yemen, on 17 December 2009, killed 41 civilians, including 21 children. On the 12 December 2013, another drone strike in Yemen allegedly killed a group of 15 civilians participating in a funeral. During the month of October 2014 and despite the fact that the number of targeted killings had decisively decreased in 2014, it is reported that the CIA launched nine strikes in Pakistan, killing between 29 and 49 people; in the same period in Yemen, between 4 and 34 people (among whom were allegedly up to 20 civilians) were killed by at least one confirmed drone strike by the United States.[19]

4. Targeted Killings by Drones as a Modern Warfare State-sponsored Practice That Requires Investigation

As the number of publications on this issue in the last few years shows, targeted killings, in particular by armed drones, are in the eye of the storm. But drones are in principle not different from other targeted killings methods: a missile fired from a drone is no different from any other commonly used weapon. The basic question is the same for each weapon: whether its use complies with International Humanitarian Law (IHL) and human rights principles.[20] It is not the technology as such that is the problem. As it has been correctly noted, in the last century there was more dramatic technological advancement with regard to weapons and warfare conduct than drones: let us think of nuclear weapons.[21] Drones as a matter of fact are still controlled by individuals who are located in a chain of command and who are responsible for the consequences of the operations: indeed, these machines are not fully autonomous killer weapons.

Surely, there are clear reasons for the escalation in the use of armed drones over the last years: militarily speaking, drones are economically advantageous and can present some advantages in terms of minimizing collateral damages.[22] But most importantly, they do not put at risk the lives of the soldiers/pilots. As has been noted by Gregoire Chamayou in his very interesting book on the philosophical foundations of drones, this is consistent with a sort of new principle, that of the "immunity of the combatant", which has emerged in recent times and according to which the lives of the soldiers are more important than those of the foreign (enemy) civilians.[23] Notably, this is in contradiction with well established principles and rules governing the conduct of hostilities.

However, this is exactly the heart of the matter: because drones make it easier to kill without risk to a State's forces, policy makers and commanders are tempted to interpret the legal limitations on who can be killed, and under what circumstances, too expansively. Thus, even if in principle drones are not different from other killing techniques, the point is that drones made targeted killings more attractive for policy makers and commanders, causing an expansion in this practice.[24] Given these premises, some commentators also highlight the existence of specific negative features of the drones: the risk of developing what has been called a 'Playstation mentality', due to the remote control through the screen and the zero risk to the forces operating the drones,[25] as well as the fact that drones are not able to conduct a capture/kill operation and can be tasked only with the killing of the target.[26]

It is not surprising that under the Obama administration, in years when the drone technology was already easily accessible, the practice to eliminate targets has escalated and expanded. What is particularly worrying is the expansion of the targeted killings practice outside the battlefield. In fact, while some targeted killings have been carried out within armed conflicts, as in the context of the wars in Afghanistan and Iraq, many have taken place outside the context of an armed conflict. U.S. strikes have been carried out in Yemen, Somalia, Sudan, Pakistan, and the Philippines. The Obama administration has already reportedly launched eight times as many strikes as the Bush administration, in fewer years in office.[27] In Pakistan alone, out of more than 400 strikes since 2004, 350 have been carried out under Obama, killing between 2.383 and 3.858 individuals (not to mention the injured ones).[28]

Some scholars have noted that there was a shift in the recent counterterrorism strategy of the United States, from the capture to the direct killing of the suspects. Indeed if Bush's national security doctrine was defined by a paradigm of capture (often in the form of illegal renditions), detention, and torture of enemies/suspected terrorists, Obama's aims directly at their elimination.[29] Yet, whilst human rights advocates were successful in building strong opposition against and effective legal challenges to torture programs, such a coherent and effective opposition has still to emerge against the use of drones for targeted killing practices, leaving a massive accountability vacuum that has enabled States to pursue this practice largely unopposed.

The UN Special Rapporteur on extrajudicial, summary, or arbitrary executions repeatedly asked the United States to explain how they justify the use of drones to target and kill individuals. Governments "should specify the basis for decisions to kill rather than capture particular individuals . . . ensure in advance of drone killings that they comply with international law, and make . . . public the number of civilians collaterally killed as a result of drone attacks, and the measures in place to prevent such casualties".[30] As it was effectively summarized by law professor and former Defense Department advisor, Rosa Brooks, with regard to the U.S. targeted killings program: "Right now we have the executive branch making the claim that it has the right to kill anyone, anywhere on earth at any time for secret reasons based on secret evidence in a secret process undertaken by unidentified officials".[31]

Given these circumstances, growing worldwide concern is not surprising. This is also testified to by the fact that, both at the national and the international levels, a number of expert panels, committees, audiences, and sessions are organised specifically to address drone strikes policy, including at the European level.[32] UN agencies are also intensively dealing with this issue: Beside the already mentioned Alston 2010 Report, the UN Special Rapporteur on Counter Terrorism and Human Rights, Ben Emmerson, was tasked with a wide range investigation regarding the civilian impact and human rights implications of the use of drones and other forms of targeted killings (Emmerson 2013). Following an interim 2013 Report, his final report, presented to the Human Rights Council (HRC) in March 2014, examined 37 drone strikes involving reports of civilian casualties: they included strikes committed by the United States, United Kingdom, and Israel in Pakistan, Afghanistan, Yemen, Somalia, and Gaza (Palestine) (Emmerson 2014).[33] According to the Special Rapporteur, these strikes require a legal duty on the relevant States to provide a public explanation of the circumstances and a justification for the use of lethal force. It is not a matter of political demand; it is a matter of accountability.[34] As the Special Rapporteur underscored, the States responsible for these strikes are under a present and continuing obligation to disclose the results of their own fact-finding inquires or explain why it is that no such inquiries have so far taken place.[35]

The Report also stresses once again the "legal uncertainty" as to the principles of international law applicable to the use of deadly force in counterterrorism operations, and the differences of State practice in this regard. These circumstances not only fail to provide adequate protection for the right to life but also pose a threat to the international legal order and can undermine international peace and security.[36] In fact, what the UN is aiming to achieve is a *multilateral legal framework* for drone strikes and other counter-insurgency operations. Towards this aim, in April 2014 the UN HRC passed a resolution[37] pursuant to which it convened an important panel discussion in Geneva on "the use of remotely piloted aircraft or armed drones in counterterrorism and military operations in accordance with international law, including international human rights and humanitarian law", with the participation of the relevant UN Special Rapporteurs and experts in the field.[38] Much is still to be done, and there are huge obstacles to overcome before achieving some agreement on the legal framework applicable to lethal operations by means of drones,[39] but it looks as though some first steps have been lately undertaken in this direction.

5. Unlawful Targeted Killings as Violations of the Right to Life and Possible War Crimes: Responsibility Issues

The issue of the permissibility under international law of the use of drones and of the targeted killing practice is an extremely complex one. What is permitted during an armed conflict (under IHL) is not always lawful outside of an armed conflict (under the law enforcement regime and human rights law). Moreover, IHL rules regulating the conduct of hostilities differ depending on the nature of the conflict.[40] As we shall see, the clash between humanitarian law and human rights law becomes apparent precisely with regard to the targeted killings issue.

As a warfare practice, targeted killings represent a distortion to the extent that the typical collective dimension of the 'enemy' is replaced by a sort of individualization and personalization of it. In this sense, targeted killings do not comply with the basic logic and underlying principles regarding the conduct of hostilities, which is not to kill the enemy but to defeat them. As the ICRC stresses, the use of lethal force should be permitted only to achieve that purpose, while the focus on individualised warfare tends to distort our understanding of the inherent purpose of the conduct of the hostility.[41]

Moreover, there are many difficulties in applying legal categories to targeted killings in the context of contemporary asymmetric conflicts, a structural characteristic of modern-day warfare,[42] due to the unclear boundaries of such conflicts and the related uncertainty as to the legal regimes applicable thereto. In fact, the very pillars upon which the *ius in bello* is based – such as the existence of an armed conflict; its nature, whether international or non-international; the distinction between combatants and civilians – are ever changing in contemporary conflicts. As a consequence, one of the basic principles of IHL, which is to avoid victims among the civilian population, is at risk because the boundary that shall distinguish the military objects from the protected civilian ones often blurs and becomes confused. Moreover, military imbalances and the non-recognition of the 'enemy' as 'a legitimate party' can push both parties to circumventing accepted rules of war: thus direct attacks against civilians, hostage taking, and the use of human shields have seen a revival in recent conflicts. On the other side, militarily superior parties, in order to strike at their 'amorphous enemy', have had to recourse to indiscriminate attacks, illegal interrogation practices, and renditions, as well as to targeted killings.[43]

Let us think, for instance, of the so-called war on terror that the United States declared after September 11, 2001 against al-Qaeda: despite its improper name, the majority view holds that under international law the 'war on terror' does not actually amount to an armed conflict as a whole: thus the operations conducted by the United States and its allied forces outside the battlefield (for instance in Yemen or Pakistan) shall be governed by the more restrictive law enforcement paradigm (and not by IHL).[44]

It shall be noted in this regard that the IHL regime is more tolerant than the law enforcement one (applicable outside of an armed conflict). Indeed, resort by States to targeted killings 'outside the battlefield'[45] is permitted only in very exceptional circumstances. In particular, in order to be lawful under the normative paradigm of law enforcement, a targeted killing must (a) have a sufficient legal basis in domestic law in accordance with the international principles on the use of lethal force; (b) be of exclusively preventive nature (not punitive); (c) aim exclusively at protecting human life from unlawful attack; (d) be absolutely necessary for the achievement of the aim; (e) be the undesired *ultima ratio* and not the actual aim of the planned operation.[46]

Differently, those targeted killings that constitute an integral part of the hostilities and that are committed within an armed conflict are regulated according to IHL rules. IHL determines whether an individual can be subject to direct attack (i.e. is a legitimate military target) and provides the standards to be respected in conducting the attack as, for instance, those dictated by Article 57(2) of Additional Protocol I to the 1949 Geneva Conventions. If the individual is a protected person under IHL (i.e. a civilian not taking part into the hostilities), then deliberate lethal force can be used only according to the very strict standards of the law enforcement paradigm. The determination of who can be a legitimate target lies, of course, at the heart of the problem.[47] Most problems arise with regard to non-international armed conflicts. In non-international armed conflict, contrary to the international ones, there is no such thing as a 'combatant'.[48] States can attack only civilians who "directly participate in hostilities". But the problem is precisely that there is no commonly accepted definition of what constitutes direct participation in the hostilities, which has been left open to States' (obscure) interpretation.[49]

Along the same lines, it is debated also whether under common Article 3 of the Geneva Conventions (dictated for non-international armed conflicts), a State is obliged to apprehend, instead of killing, suspected terrorists/enemy fighters. In fact, in a non-international armed conflict, non-State fighters are not protected against attack when they are taking direct or active part in the hostilities. The debate, under humanitarian law, rages over what types of activities trigger this loss of immunity and whether individuals deemed to be 'enemy combatants' can become legitimate targets of military operations. In other words, the application of common Article 3 standards to these situations may not bar States from killing suspected terrorists, even if no previous attempt was made to arrest or detain them. However, it has been suggested that reading IHL together with human rights law provisions would produce a rule that does require States to prefer the apprehension of enemies/suspected terrorists over killing them.[50] Be that as it may, it is undisputable that international human rights principles do not cease to apply in time of war, as long recognised by the International Court of Justice (ICJ); this implies that we have to deal with a mixed system composed of IHL and human rights principles.

In fact, the most progressive doctrine considers that some fundamental human rights principles, which place at the forefront the protection of the right to life and hold recourse to lethal force as legitimate only in the absence of less severe alternatives (*ultima ratio*), shall find application also in time of war. Thus, the following criteria have been outlined, *inter alia*, in order for a targeted killing to be lawful during an armed conflict: (a) it needs to be committed as part of the hostilities; (b) it is likely to contribute effectively to the achievement of a concrete and direct military advantage; (c) there is no non-lethal alternative to achieving the military advantage; (d) it is directed against a non-protected person; (e) it is not disproportionate (in relation to the concrete and direct military advantage anticipated); (f) it is planned and conducted so as to avoid erroneous targeting and to minimise harming civilian objects; (g) it may be suspended if the targeted individual surrenders.

A thorough legal analysis would show that most of the targeted killings conducted outside the battlefield (and therefore regulated by the law enforcement paradigm) constitute premeditated deprivations of life that can be only rarely justified by unprovoked, absolute, and immediate necessity within the meaning of self-defence.[51] Moreover, it appears that even among those strikes conducted in the context of an armed conflict (and therefore regulated by the mixed IHL/human rights paradigm), many would be unlawful because they are disproportionate. According to Article 57 of Additional Protocol I, those who plan or decide upon an attack shall:

a Do everything feasible to verify that the objectives to be attacked are neither civilians nor civilian objects . . .;
b Take all feasible precautions in the choice of means and methods of attack with a view to avoiding, and in any event to minimizing, incidental loss of civilian life, injury to civilians and damage to civilian objects;
c Refrain from deciding to launch any attack which may be expected to cause incidental loss of civilian life, injury to civilians, damage to civilian objects, or a combination thereof, which would be excessive in relation to the concrete and direct military advantage anticipated.

Moreover, "the attack shall be cancelled or suspended if it becomes apparent that the objective is not a military one or is subject to special protection or that the attack may be expected to cause incidental loss of civilian life, injury to civilians, damage to civilian objects, or a combination thereof, which would be excessive in relation to the concrete and direct military advantage anticipated". And "effective advance warning shall be given of attacks, which may affect the civilian population, unless circumstances do not permit".

The violation of these rules and principles can give rise to the commission of war crimes, in the form of direct attacks against civilians or disproportionate attacks.[52] Outside the context of an armed conflict and when committed as part of a widespread or systematic attack against the civilian population, the same conduct could constitute forms of crimes against humanity.[53]

6. The Duty of States to Protect the Right to Life and Investigate Its Violations

It is worrisome that, in the face of hundreds of cases of alleged arbitrary deprivation of life, those States that use drone strikes have refused so far to conduct any serious investigation. Indeed, the major "accountability vacuum" reported by Alston back in 2010[54] is only growing bigger in recent years. It shall be recalled that, with the sole exception of operations directed against persons of undisputed combat status, States have an international obligation to conduct an independent and effective investigation into the lawfulness of each case of targeted killing attributable to them.[55] Indeed, "in any case in which there have been, or appear to have been, civilian casualties that were not anticipated when the attack was planned, the State responsible is under a obligation to conduct a prompt, independent and impartial fact-finding inquiry and to provide a detailed public explanation of the results".[56] Notably, this obligation is triggered whenever there is a "*plausible* indication from any *apparently reliable* source" of unintended civilian casualties, including when the "facts are *unclear* or the information is *partial or circumstantial*".[57]

In fact, contrary to military necessity, absolute necessity cannot be assessed on a collective basis, i.e. for a whole group or category of persons (for instance, the "al-Qaeda terrorist groups"). There is a broad misconception and misleading rhetoric as to the fact that suspected terrorists would be always legitimate military targets within an amorphous 'war on terrorism' (which, as we have seen, does not amount as a whole to an armed conflict under IHL). On the contrary, according to legal standards, outside the conduct of hostilities no person can be lawfully deprived of their life based on criteria other than their conduct at the moment of the use of lethal force: definitely not based on their past behaviour (except death penalty, in those systems that still have it) or their belonging to a group or category.[58]

In sum, what emerges clearly from this somehow limited analysis is that (apart from exceptional cases) the permissibility of the use of lethal force in the drones strikes/targeted killing operations can only and must be determined on a case-by-case basis.[59] States are indeed under an obligation to investigate. In fact, the human rights obligation to investigate alleged violations of the right to life promptly, thoroughly, and effectively through independent and impartial bodies does not cease to apply even during armed conflicts.[60] Not surprisingly, however, States are reluctant to acknowledge their obligation to respect the legal standards pertaining to the right to life, particularly when targeting individuals outside their territorial jurisdiction, and have a correspondent counter-interest to establishing prompt, transparent, and genuine investigations. Since targeted killings are normally conducted against non-citizens, i.e. foreigners not covered by the domestic constitutional right of protection of their right to life, it is very difficult to activate judicial procedures in order to establish the responsibility of the actors involved. At the same time, as we will see in the next paragraph, even when, in exceptional cases, it is possible to bring claims in front of a judicial authority, political concerns and vague national security arguments invoked by the governments have so far impeded the judges from adjudicating the issue.

7. The Difficult Assessment of Liability for Targeted Killings Operations before Domestic Courts

In order to show the difficulties related to the ascertainment of responsibility regarding the State-sanctioned targeted killings practice, it is interesting to have a look at some cases that are exemplificative of the previously mentioned dynamics. The first one is a recent case regarding the killing in 2011 of American citizens in

Yemen by U.S. drone strikes, which is particularly important because it reached – although unsuccessfully so far – the docket of a judge.

7.1 The Al-Aulaqi case

The opportunity to bring a targeted killing case before the U.S. judiciary appeared in conjunction with the targeted killing of some American citizens by U.S. drone strikes. As the judge assigned to the case put it, "A foreign national targeted by the United States in a foreign place may not have . . . constitutional rights, a U.S. citizen does." On 30 September 2011, a U.S. drone strike killed Anwar Al-Aulaqi, an American citizen who had been placed on government "kill-list" over a year before, and Samir Khan, another American citizen. Two weeks later, on 14 October, another U.S. drone strike in Yemen killed Anwar Al-Aulaqi's son, 16-year-old Abdulrahman Al-Aulaqi, as he was eating dinner with his teenage cousin at an open air restaurant. Several other people were also killed in the two strikes.

The case *Al-Aulaqi v. Panetta et al.* was filed on 18 July 2012; it was directed against senior CIA and military officials, arguing that the killing of three American citizens by their own government by means of drone strikes was in violation of national and international law. The killings, which took place in Yemen, a country not alleged to be at war with the United States, are considered to have been carried out outside of an armed conflict; thus the law enforcement paradigm, and in the first place the U.S. Constitution and International Human Rights Law, shall govern the operation. Specifically, the claimants denounced the violation of the Fifth Amendment to the Constitution, which provides that no person shall be deprived of life without due process of law, and the Fourth Amendment, which prohibits the excessive use of force. International Human Rights Law similarly protects against the arbitrary deprivation of life and generally prohibits killings by States without a due legal process. In the absence of due process, according to the United States and international human rights standard, the only permissible deprivation of life is in case of an imminent threat of death or serious physical harm and then only if lethal force is the last resort.

The case was brought by the Centre for Constitutional Rights (CCR) and American Civil Liberties Union (ACLU) against four senior U.S. officials responsible for authorizing and directing the killings by the CIA and the Joint Special Operations Command (JSOC).[61] As argued by the claimants, these targeted killings relied on vague legal standards, a closed executive process, and evidence never presented to any court. Indeed Anwar Al-Aulaqi was designated for death by the executive branch – in other words, he was put on the infamous "Kill-list" checked personally by President Obama – without the protection of any judicial review. It is indeed a troubling question that was raised (but not answered) by the District Court Judge who decided on the first petition that sought to challenge the inclusion of Al-Aulaqi in the Kill-list:

> How is it that judicial approval is required when the United States decides to target a US citizen overseas for electronic surveillance, but that, according to defendants [i.e. President Obama, then CIA Director Leon Panetta and then

> Defense Secretary Robert Gates] judicial scrutiny is prohibited when the US decides to target a US citizen overseas for death?[62]

The Department of Justice, representing the defendants, has moved to dismiss the case, arguing that there is no role for the judiciary in reviewing the claims because they raise "political questions" and "national security concerns" and that the defendants should be immune. The Justice Department thus has continued to argue that national security concerns bar any judicial review of any government's action.

On 4 April 2014, the U.S. District Judge Rosemary M. Collyer dismissed the case.[63] The judge, while recognizing that "the executive is not an effective check on the executive" and while rejecting the government's argument that the case presented a "political question" that prevented the judiciary from hearing the case, nevertheless dismissed the Al-Aulaqi case.

The dismissal was based on two sets of spurious arguments: on the one hand, regarding the killing of Anwar Al-Aulaqi, the judge effectively accepted the governments' allegation that he was a grave threat and that it was not possible to neutralize him through non-lethal means, in a way condemning him posthumously. On the other hand, with regard to the other two victims, because Samir Khan and 16-year-old Abdulrahman Al-Aulaqi had not been (allegedly) targeted by the U.S. government and they were killed only as collateral damages, or mistakes, the Court found that their constitutional rights were not violated. In the end, as one of the lawyers representing the victims put it, "it seems that there is no remedy if the government intended to kill you, and no remedy if it didn't".[64]

Given the closing of the case, the very hard question raised by the judge herself remains unanswered. "At what point could a President and only a couple of members in the Senate . . ., at what point can they not decide that an American citizen overseas is a danger to the country and therefore, without process of law be killed?" For sure, the Court's decision is deeply troubling, and its effects raise serious questions of constitutionality, to the extent that violations of a U.S. citizen's right to life cannot be heard in a federal court.

7.2 Two Very Recent Controversial Cases: The First Cracks in the Wall?

Notably, a recent case regarding another American citizen who was on the CIA and Pentagon's Kill-list, which apparently resembled very much the case of Anwar Al-Aulaqi, had quite the opposite outcome. The case involves Mohanad Mahmoud Al Farekh, a young American citizen born in Texas and raised in Canada, who is believed to have been living in Pakistan since 2007. For his alleged affiliation with terrorist organizations, the CIA and the Defence Department in 2013 requested approval to put him on the kill-list, claiming that: (a) he posed a continuing and imminent threat to the American people; (b) no other governments were capable of effectively addressing the threat; (c) capture was infeasible; and (d) there was a near certainty that no civilians would be killed or injured.[65] It is reported by the media that it was only thanks to the intervention of the Attorney

General and other Justice Department officials that Al Farekh's inclusion in the kill-list in the end was not approved.[66]

Nevertheless, a warrant of arrest was issued by the District Court of New York,[67] and Al Farekh – instead of being summarily killed by a drone strike – was finally arrested by Pakistani forces and surrendered to the United States, where he was indicted on criminal charges. According to the official press release by the U.S. Department of Justice (DoJ), Al Farekh is alleged to have conspired with others to provide material support to terrorists. He "allegedly provided material support to terrorists with every intention of becoming a martyr".[68] If convicted, the defendant faces a maximum sentence of 15 years' imprisonment. The DoJ communication in the end rightly explicates that "the charges in the complaint are merely allegations and the defendant is presumed innocent unless and until proven guilty". Thus the (obvious) question: what about the presumption of innocence of all those individuals killed by targeted killings operations? Moreover, as *The New Yorker* journalist rightly notes, how imminent, or even just urgent, could the threat posed by Al Farekh really have been, if two years later it was possible to arrest him?[69] Would the Justice Department have raised doubts about an authorization to kill him if Al Farekh had not been a U.S. citizen?[70] In this regard for sure, the impact of the Al-Aulaqi case – which had quite an echo in the media and public opinion, even if only indirectly – played a big role in the decision not to put another American citizen on the Kill-list.

Another huge scandal emerged very recently, when the U.S. administration had to admit that two civilians, the U.S. citizen Warren Weinstein and the Italian Giovanni Lo Porto, were killed by mistake by a U.S. drone strike in Pakistan in January 2015. Obama publicly apologized for the incident and ensured that a "thorough independent review" will be conducted. Moreover, the U.S. president publicly spoke about compensation to the victims' families.[71] Clearly, these unprecedented words can be already seen as a turning point, a positive step in the direction of ensuring accountability for the violations of fundamental human rights through the drones/targeted killings program. As the major human rights organizations requested in an open letter to the U.S. president on May 2015, it is urgent that the U.S. administration adopt the same approach to all other U.S. strikes where civilians have been killed or injured, regardless of their nationalities.[72] In particular, it is necessary that the government establish a systematic and transparent mechanism for reviewing and investigating the strikes, that such investigations are made public, and that appropriate redress is provided to civilian victims.

Maybe these cases are just the first cracks in the wall, but for sure something is moving. As already happened with regard to the opposition to torture practices, it is becoming more and more evident that the counterterrorism rationale cannot justify the blatant or the secret violation of the fundamental human rights of the individuals.

Notes

1 See the data, which are monthly adjourned, by the Bureau of Investigative Journalism. Available online at http://www.thebureauinvestigates.com.

2 See *infra*, Section 5.

3 Among the several works that have been published in the past years by international and criminal lawyers on this specific issue, especially in the United States, see the collective volumes by Cohn (ed) (2015); Knuckey (ed) (2014); Finkelstein, Ohlin, and Altman (eds) (2012); and the fundamental, in-depth study by Melzer (2008). For an overview of the most recent works, see the list, compiled by the Bard College's Center for the Study of the Drone, of the best "Drone Books" of 2014. Available online at http://dronecenter.bard.edu/a-winter-reading-guide-to-drones/.
4 See Alston (2010) Report, Summary and paras. 1–3 (emphasis added).
5 See Melzer (2008: 239).
6 Alston (2010) Report, para. 7.
7 Alston (2010) Report, para. 8.
8 However, the identification of the individuals in advance does not seem to be anymore a criterion to define targeted killings. The signature strikes in fact are precisely conducted against non-identified people, see (Heller 2013: 89). Better then to speak of "individually selected persons", see (Melzer 2008: 3).
9 See Alston (2010) Report.
10 As it was brilliantly noted, "the rhetoric of self-defense was one of the most frequently abused concepts of the twentieth century" (Fletcher and Ohlin 2008: 3). As the authors affirm, part of the confusion around the concept stems from a basic misunderstanding about the distinct concepts of self-defence under respectively international law and (international) criminal law. One thing is in fact to legitimize the use of armed force by a State resorting to the self-defence argument (under Article 51 of the UN Charter); another is to invoke self-defence as an excuse for the (otherwise) criminal behaviour of individuals, for instance those in position of authority and command, with regard to the killing of criminals/enemies/terrorists. See again Fletcher and Ohlin (2008: 8–11 and 30–106).
11 Contra, see Kasher (in this volume). However, it shall be noted that Kasher's argumentation seems to be based on an erroneous concept of a "State's duty to self-defence", which in fact does not exist as such under the international legal order. One thing is the *right* to self-defence; a different one is the State's *duty* to protect its citizens from threats. Moreover, it shall not be forgotten that States also have a positive duty to protect the civilians (and not only their citizens) under their control during an armed conflict; see Rule 22 of the Customary IHL Rules by the ICRC. Available online at https://www.icrc.org/customary-ihl/eng/docs/v1_rul_rule22.
12 See Alston (2010) Report, para. 13.
13 David (2003: 111–126). See also Melzer (2008: 27 ss).
14 See the report by Dr. Atef Abu Saif (2014).
15 On the contrary, many targeted killings conducted by Israel in Gaza are all but surgical, some of them causing dozen of innocent civilian victims. An exemplar in this sense is the case of Salah Shehadeh, who was killed on 22 July 2002 by a one-tonne bomb dropped at night on his building in a highly populated area of Gaza city, causing the death of 14 innocent civilians, including eight children and injuring another 150 people. (See Weill 2009.) Notably, the targeted killings practice has come under the scrutiny of the Supreme Court in Israel, but, although worried by this practice, the Court failed to declare its unlawfulness in principle, finding that the assessment must be conducted on a "case by case basis"; for a critique, see Eichensehr (2007).
16 See Emmerson (2013) Report, paras. 31, 49.
17 See Alston (2010) Report, para. 23.
18 See Ambos and Alkatout (2012: 341–366).
19 Bureau of Investigative Journalism, Monthly Update, October 2014. Available online at http://www.thebureauinvestigates.com.
20 See Alston (2010) Report.
21 Cf. Boyle (2015: 106).
22 See the Emmerson (2013) Report, paras. 25–28.

23 Chamayou (2014: 129ff.).
24 Cf. also Boyle (2015: 121).
25 But see Derek (2011: 197ff.).
26 See also Koch (2014) and Chapter 4 in this volume.
27 See CCR factsheet. Available online at www.ccrjustice.org.
28 Bureau of Investigative Journalism, Monthly Update, October 2014. Available online at http://www.thebureauinvestigates.com.
29 See Hajjar (2012). Available online at http://www.merip.org/mer/mer264/anatomy-us-targeted-killing-policy.
30 Philip Alston, Mission to the United States of America, Addendum to UN Human Right Council Report, 28 May 2009. http://www2.ohchr.org/english/bodies/hrcouncil/docs/11session/A.HRC.11.2.Add.5.pdf.
31 Professor Rosa Brooks, Georgetown University Law Center, during the first public U.S. Senate Judiciary hearing on the use of drones, 24 April 2013.
32 See European Parliament Resolution on the Use of Armed Drones of 25 February 2014 [2014/2567 (RSP)], calling, *inter alia*, upon the High Representative for Foreign Affairs and Security Policy, the member States, and the Council of the European Union to "oppose and ban the practice of extrajudicial targeted killings" and urging the Council to adopt a European Union common position on the use of armed drones.
33 See Emmerson (2014) Report.
34 See *infra*, para. 6.
35 See Emmerson (2014) Report, para. 36.
36 Ibid., para. 70 ss.
37 A/HRC/Res/25/22 of 15 April 2014.
38 See the summary report of the Panel, held on 22 September 2014 at the HRC in Geneva. Available online at http://www.unog.ch/unog/website/news_media.nsf/%28httpNewsByYear_en%29/BCE56ED914A46D40C1257D5B0038393F?OpenDocument.
39 To have an idea of the difficulties, see the list of open questions in the (Emmerson 2014) Report, paras. 70–71; on the issue more thoroughly *infra*, para. 6.
40 For a thorough analysis see Kretzmer (2005: 171–212) and, again, the fundamental study by Melzer (2008). Notably, both authors, but in particular the latter, refute the position that the current legal framework had become inadequate to regulate States' resort to lethal force; on the contrary, although extremely complex to analyse, the legal framework composed by both IHL and HR law would provide clear and unambiguous standards as to State-sponsored targeted killing operations; rather, it is the State practice that, disregarding the clearly applicable legal standards, is at the heart of the problem.
41 See Melzer (2008: 241 ss); see also Koch in this volume.
42 See Geiss (2006: 757 ss).
43 Ibid.
44 Melzer (2008: 262 ss); see also Alston (2010) Report, para. 53.
45 On the difficulties of defining the geographical boundaries of the 'battlefield' and consequently the legal regime applicable to the use of contemporary asymmetrical conflicts, see Lubell and Derejko (2013: 65 ss).
46 Melzer (2008: 222–239).
47 Alston (2010) Report, para. 57.
48 See Sassoli and Olson (2008: 599ff.).
49 In 2009, the ICRC issued its Interpretive Guidance on "direct participation in the hostilities". In non-international armed conflict, according to the ICRC, civilians who participate directly in hostilities and are members of an armed group who have a "continuous combat function" may be targeted at all times and in all places (ICRC Guidance, at 66). Whether this interpretation of the "continuous combat function" reflects in fact international customary law is still disputed: see Emmerson (2014) Report, para. 71(f).

50 See in this sense (Melzer 2008: 382ff.).
51 Ibid. *Contra*, see Kasher in this volume; as previously noted, however, the author's interpretation of the self-defence argument is based on very different premises.
52 See Alston (2010) Report, para. 72. See in particular Art. 8(2)(b)(i) and Art. 8(2)(b)(iv) of the ICC Statute.
53 See Art. 7(1)(a) and Art. 7(1)(g) of the ICC Statute.
54 See supra, para. 2.
55 See the European Parliament Resolution on the Use of Armed Drones of 25 February 2014.
56 See Emmerson (2014) Report, para. 32, and Emmerson (2013) Report, para. 78.
57 Ibid. (emphasis added).
58 See the in-depth analysis of Melzer (2008: 83 ss).
59 See Israel High Court of Justice, *The Public Committee Against Torture et al. v. The Government of Israel, et al.* HCJ 769/02, Judgment of 14 December 2006 (Targeted Killings Decision).
60 See Alston (2010) Report, para. H.
61 Notably, the four defendants are some of the highest-ranking military and civilian leaders of the country: former CIA Director David Petraeus, Secretary of Defense Leon Panetta, Commander of the Special Operations Command William McCraven, and Joint Special Operations Commander Joseph Votel.
62 See CCR factsheet and transcripts of the case. Available online at www.ccrjustice.org.
63 District Judge for the District of Columbia, *Nasser Al-Aulaqi v. Leon Panetta et al.*, Civil Action No. 12–1192 (RMC).
64 See CCR Press Release, "Court Dismisses Lawsuit Challenging U.S. Drone Killings of Three Americans", 4 April 2014. Available online at https://www.aclu.org/news/court-dismisses-lawsuit-challenging-us-drone-killings-three-americans.
65 These were the conditions set by the "Guidelines" adopted by the White House, as acknowledged by President Obama in his Remarks at the National Defense University, Washington, on 23 May 2013.
66 See Cole (2015).
67 *United States of America against Muhanad Mahmoud Al Farekh*, Eastern District of New York, 2 April 2015.
68 U.S. Department of Justice (2015).
69 Cf. Cole (2015).
70 Ibid.
71 The video and full text of Obama's remarks on the killing of Weinstein and Lo Porto have been published by *The Washington Post* on 23 April 2015 and is available on the Journal's website. https://www.washingtonpost.com/world/national-security/us-operation-kills-al-qaeda-hostages-including-american/2015/04/23/8e9fcaba-e9bd-11e4-aae1-d642717d8afa_story.html.
72 Cf. the open letter [by ACLU, Amnesty International, CIVIC, Centre for Constitutional Rights, European Centre for Constitutional and Human Rights, Human Rights Clinic (Columbia Law School), Human Rights First, Human Rights Watch, Open Society Foundations, Reprieve], regarding "Public Acknowledgment and Investigations of U.S. 'Targeted Killings' and Drone Strikes, 13 May, 2015". The letter also includes an attachment with a list of 10 examples of strikes that require urgent investigation because of serious allegations of civilian casualties.

References

Alston, P. 2010. *Study on Targeted Killing, Addendum to the Report of the Special Rapporteur on Extrajudicial, Summary or Arbitrary Executions*, 28 May. A/HRC/14/24/Add.6 (Alston 2010 Report). Available online at http://www2.ohchr.org/english/bodies/hrcouncil/docs/14session/A.HRC.14.24.Add6.pdf.

Ambos K. and Alkatout J., 2012. Has "Justice Been Done"? The Legality of Bin Laden's Killing under International Law, *Israel Law Review* 45(2): 341–366.

Boyle M.J., 2015. The Legal and Ethical Implications of Drone Warfare, *The International Journal of Human Rights* 19(2): 105–126.

Chamayou G., 2014. *Teoria del Drone*, Derive e Approdi.

Cohn M. (ed.), 2015. *Drones and Targeted Killing: Legal, Moral, and Geopolitical Issues*, Olive Branch Press.

Cole D., 2015. *Targeted Killing: The New Questions*, *The New Yorker Review of Books blog*, 14 April. Available online at www.nybooks.com/blogs/nyrblog/2015/apr/14/targeted-killing-new-questions.

David R., 2003. Israel's Policy of Targeted Killings, *Ethics and International Affairs* 17(1): 111–126.

Derek G., 2011. From a View to a Kill: Drones and Late Modern War, *Theory Culture Society* 28(7–8): 188–215.

Eichensehr K.E., 2007. On Target? The Israeli Supreme Court and the Expansion of Targeted Killings, *The Yale Law Journal* 116(8): 1873–1881.

Emmerson, B. 2013. *2013 Interim Report to the UN General Assembly of the Special Rapporteur on the Promotion and Protection of Human Rights and Fundamental Freedoms while Countering Terrorism*, 18 September, A/68/389 (Emmerson 2013 Report). Available online at https://www.google.com/search?q=%22Study+on+Targeted+Killing%2C+Addendum+to+the+Report+of+the+Special+Rapporteur+on+Extrajudicial%2C+Summary+or+Arbitrary+Executions%22&ie=utf-8&oe=utf-8#q=%22Interim+Report+to+the+UN+General+Assembly+of+the+Special+Rapporteur+on+the+Promotion+and+Protection+of+Human+Rights+and+Fundamental+Freedoms+while+Countering+Terrorism%22%2Bemmerson.

Emmerson, B. 2014. *Report of the Special Rapporteur on the Promotion and Protection of Human Rights and Fundamental Freedoms while Countering Terrorism*, 11 March, A/HRC/25/59 (Emmerson 2014 Report).

European Parliament Resolution on the Use of Armed Drones, 2014. 25 February [2014/2567 (RSP)].

Finkelstein C., Ohlin J.D., and Altman A. (eds), 2012. *Targeted Killings: Law and Morality in an Asymmetrical World*, Oxford University Press.

Fletcher G., and Ohlin J.D., 2008. *Defending Humanity*, Oxford University Press.

Geiss R., 2006. Asymmetric Conflict Structures, *International Review of the Red Cross* 88(864): 757–777.

Hajjar L., 2012. Anatomy of the US Targeted Killing Policy, Middle East Reasearch and Information Project, *MER* 42(264). Available online at http://www.merip.org/mer/mer264/anatomy-us-targeted-killing-policy.

Heller K.J., 2013. "One Hell of a Killing Machine". Signature Strikes and International Law, *Journal of International Criminal Justice*, 11: 89–119.

Knuckey S. (ed), 2014. *Drones and Targeted Killings. Ethics, Law, Politics*, Idea.

Kretzmer D., 2005. Targeted Killing of Suspected Terrorists: Extra-Judicial Executions or Legitimate Means of Defence? *European Journal of International Law* 16(2): 171–212.

Lubell N., and Derejko N., 2013. A Global Battlefield? Drones and the Geographical Scope of Armed Conflict, *Journal of International Criminal Justice* 11(1): 65–88.

Melzer N., 2008. *Targeted Killing in International Law*, Oxford University Press.

Saif, Atef Abu. 2014. *Sleepless in Gaza. Israeli Drone War on the Gaza Strip*, 2014. Rosa Luxemburg Stiftung Bureau of Investigative Journalism, Monthly Update. Available online at http://www.rosa-luxemburg.com/wp-content/uploads/2014/02/Sleepless-in-Gaza-by-Atef-Abu-Saif-RLS-Palestine.pdf.

Sassoli M., and Olson L.M., 2008. The Relationship between International Humanitarian Law and Human Rights Law Where It Matters: Admissible Killing and Internment of Fighters in Non-international Armed Conflicts, *International Review of the Red Cross* 90(871): 599–627.

U.S. Department of Justice, Office of Public Affairs, 2015. *American Citizen Charged with Conspiring to Provide Material Support to Terrorists*, 2 April. Available online at www.justice.gov/opa/pr/american-citizen-charged-conspiring-provide-material-support-terrorists.

Weill S., 2009. The Targeted Killing of Salah Shehadeh. From Gaza to Madrid, *Journal of International Criminal Justice* 7(3): 617–631.

Part II

State Responsibility and the Use of Drones

3 Autonomous Killer Robots are Probably Good News[1]

Vincent C. Müller

1. Introduction

Will future lethal autonomous weapon systems (LAWS), or 'killer robots', be a threat to humanity? The European Parliament has called for a moratorium or ban of LAWS; the Contracting Parties to the Geneva Convention at the United Nations are presently discussing such a ban, which is supported by the great majority of writers and campaigners on the issue. However, the main arguments in favour of a ban are unsound. LAWS do not support extrajudicial killings, they do not take responsibility away from humans; in fact, they increase the ability to hold humans accountable for war crimes. Using LAWS in war would probably reduce human suffering overall. Finally, the availability of LAWS would probably not increase the probability of war or other lethal conflict – especially as compared to extant remote-controlled weapons. The widespread fear of killer robots is unfounded: they are probably good news.

2. Definition

The autonomous robotic systems that are the subject of this chapter would be able to select and attack targets without intervention by a human operator. While the initial command to attack would be given by a human, the robot then has a degree of autonomous 'choice' for its actions.

Some examples of systems that fall under the scope of this chapter are automated radar-guided gun systems to defend ships that have been in use since the 1970s (e.g. the U.S. Phalanx Close-in Weapon System). Current versions can autonomously identify and attack incoming missiles, rockets, artillery fire, aircraft, and surface vessels according to criteria set by the human operator. Similar systems exist for tanks, e.g. the Russian Drozd (1977–1982) and now Arena or the German AWiSS/AVePS (Active Vehicle Protection System) by Diehl, which has a reaction time below 400 milliseconds. The main reason for not having a 'human in the loop' in these systems is sheer speed.

Systems with a higher degree of autonomy would include drones: for example, "[t]he United Kingdom Taranis jet-propelled combat drone prototype can autonomously search, identify and locate enemies but can only engage with a target when authorized by mission command. It can also defend itself against enemy aircraft" (Heyns 2013, §45). The U.S. X-47B drone can take off and land on aircraft carriers (demonstrated in 2014); it is set to be developed into an Unmanned Carrier-Launched Surveillance and Strike (UCLASS) system. It is now quite conceivable that an autonomous drone, perhaps with the size and appearance of a bird, could be commanded to locate (e.g. using a cell phone signal), pursue, and kill an individual person – rather like a 'hit man'. In a few years, some states will have the ability to deploy such a killer drone anywhere in the world, for example against someone they consider a 'terrorist'.

While autonomous air systems are fairly advanced, these kinds of systems will shortly also be available on land, on water, and under water, as well as in space. For example, the U.S. Navy has developed and tested small 'swarm boats' that can accompany a ship and protect it from small vessel attacks by detecting and swarming around such vessels – the current version has a human in the loop for weapons fire (Smalley 2014).

2.1 Problem

Robotics in warfare will be a major change, comparable to the introduction of planes or perhaps even nuclear weapons (Singer 2009b: 179, 203). One of the questions is whether the use of robotics, especially highly autonomous robotics, constitutes just one major step in the arms race or a step that introduces qualitatively new ethical concerns. Many authors and organisations have claimed that killer robots are a serious threat to humanity and should be banned, while others have said there is nothing new here (Arkin 2009, 2013). As the UN Rapporteur says in his careful and detailed report:

> Some argue that robots could never meet the requirements of international humanitarian law (IHL) or international human rights law (IHRL), and that, even if they could, as a matter of principle robots should not be granted the power to decide who should live and die. These critics call for a blanket ban on their development, production and use.
>
> (Heyns 2013, §31)

I agree that LAWS pose new problems and therefore recommend that they must meet special *technical standards* of reliability, especially with respect to 'distinction' and 'proportionality', and that their manufacturers and users abide by *legal standards* that allow allocation of responsibility.

In this chapter, I provide a general recommendation on the issue of whether killer robots should be banned, concluding that they should not. I do so by providing a concise survey of the relevant moral concerns.

2.2 Terminological Note

The UN now uses the term 'LAWS', for lethal autonomous weapon systems (Simon-Michel 2014), and I follow this, despite its unfortunate positive connotations. The weapons concerned are also often called 'LARs', for lethal autonomous robots [e.g. (Heyns 2013)]; simply 'drones' (European Parliament 2014); 'killer robots' (www.stopkillerrobots.org); 'robotic weapons' (Leveringhaus and Giacca 2014); or 'unmanned systems'. I think that the systems concerned are not just lethal in their effects but made to be lethal, that is they are weapons, so 'LARs' is insufficiently precise. 'Drones' is too narrow, since we are not talking only about flying systems, and too broad, since present drones are remote-controlled (remote-piloted). 'Unmanned' is sexist, but in any case the distinguishing feature here is not whether the human in control is actually inside the system (e.g. the plane) or controlling it from a remote location (as with current drones) but whether the system has a degree of autonomy. For this reason 'autonomous weapon system' (AWS) could be an option, but it is not used. 'Killer robots' is apt in that it implies autonomy and is not limited to air systems; while perhaps overdramatic, it makes explicit the moral issue at stake. So I use it together with 'LAWS', acknowledging the limitations of both terms.

2.3 Simple Slogans

A discussion about weapons, killing, suffering, and war often generates heated exchanges and reduction to simple slogans. Slogans may be useful campaigning tools but do not address the deep moral issues. To forestall misunderstanding, some immediate qualifications are in order. Despite the provocative title, I agree that killing and wars are great evils. More weapons are generally bad, too; they increase the probability of killing, and they are usually an inefficient use of resources. I am also deeply concerned about the current use of drones for extrajudicial killings (see Meloni and Koch in this volume, and Enemark 2014) and about the importance of civilian gun control (Müller 2015).

2.4 Structure of the Chapter

After an abbreviated look at the current technical and policy situation, I discuss the four main arguments in this debate: whether LAWS are *inherently* wrong because (a) they violate humanitarian law or (b) they make it harder or impossible to assign responsibility for killings in war; and whether *the consequences* of LAWS are good or bad in the long run – either (c) by increasing or decreasing the suffering of war or (d) by making war more or less likely. I conclude with a number of policy recommendations.

3. Current Situation

3.1 Technological

It is clear that, increasingly, autonomous weapons are coming. The first systems that make simple 'attack' decisions are already in use (see Section 2). Remote-controlled air systems have been used extensively, especially in the asymmetric Pakistan 'drone war'. Remote-controlled water, underwater, and ground systems are also in use or close to deployment (see e.g. Singer 2009a). Meanwhile, autonomous driving and sailing systems are at a high level of technological readiness, being tested outside the lab (Smalley 2014). The 2013–2038 Unmanned Systems Integrated Roadmap of the U.S. Department of Defense (U.S. Department of Defense 2013) foresees increasing levels of autonomy in air/land/sea systems in the coming 25 years. The funding for such systems is massive. In the United States only, the Department of Defense is currently spending about US$5 billion per year on 'unmanned systems' (U.S. Department of Defense 2013: 3), as well as an unknown amount of the Defense Advanced Research Projects Agency's (DARPA) US$3 billion per year budget. Further U.S. sources cannot be ruled out. While the United States is the world leader in military technology with spending on its military at about 40% of the world's defence spending (ISS 2014), other countries are certainly pursuing automated technologies, notably Russia and China.

The enabling technology for autonomous artificial intelligence (AI) is developing apace. The probability of high-level machine intelligence with full human abilities by 2040 is moving to over 50%, according to a recent survey of expert opinion (Müller and Bostrom 2015). Even if these estimates turn out to be excessively optimistic, significant autonomy levels in target selection and attack will clearly be possible in the next decade.

Killer robots are attractive to the military and thus to political funding for a number of reasons: they reduce the risk to one's own soldiers, reducing the human and political costs of war. They can be cheaper than human soldiers in the long run, not needing a salary, pension, housing, food, hospitals, and the like. They can also outperform humans and human-controlled systems, especially in terms of speed, accuracy, and ability to function without rest. They can function in environments where human remote control is not an option (e.g. underwater).

3.2 Policy

Some states, notably the United States, have developed initial policies for LAWS that include a moratorium on systems that do not have a human in the loop. However, these policies can be changed any time, at the discretion of these states. Indeed, this possibility is explicitly stated in U.S. policy: "[a]utonomous . . . weapons systems shall be designed to allow commanders and operators to exercise appropriate levels of human judgment over the use of force" (U.S. Department of Defense 2012: 2).

> Current armed unmanned systems deploy lethal force only in a fully human-operated context for engagement decisions. . . . As technology advances and more automatic features are introduced, DoD will continue to carefully consider the implications of autonomy in weapon systems to ensure safe and reliable operations and minimize the probability and consequences of failures that could lead to unintended engagements. For this reason, DoDD 3000.09, Autonomy in Weapon Systems, mandates a policy review before entering

> formal development for proposed weapon systems that would use autonomous capabilities in a novel manner.
>
> (U.S. Department of Defense 2013: 24)

The United Kingdom is working on the matter: "[t]he pace of technological development is accelerating and the UK must establish quickly a clear policy on what will constitute acceptable machine behaviour in future; there is already a significant body of scientific opinion that believes in banning autonomous weapons outright" (Ministry of Defense 2011).

An independent policy report from the University of Birmingham on "The Security Impact of Drones" came out in October 2014, and it recommends a ban on LAWS, since they "fail the test of the 'laws of humanity and the requirements of the public conscience'", and "it is unclear where responsibility would lie for any unlawful actions by weaponised robots would lie". The report also cites UK politicians saying "the UK's position on not wishing to develop such weapons is absolutely clear" (Birmingham Policy Commission 2014: 64, 65).

4. Moral Arguments

The arguments against killer robots fall into two broad categories: principled ones concerning rights and responsibility and utility considerations deriving from the likely consequences of their use.

4.1 War Crimes and International Law

Killer robots, like any weapon in war, must comply with the regulations of International Humanitarian Law in the Geneva Convention. If they cannot, they are illegal weapons, and their use in war *(in bello)* constitutes a war crime. Their use outside a war is a crime in any case (details in Emmerson 2014). There are two pertinent requirements. They must comply with the *principle of distinction*, that is have the ability to discriminate combatants from non-combatants. The legal use of LAWS would thus require a positive identification of enemy soldiers, tanks, airplanes, and the like. With present technology, this will be possible only in some situations. In particular, building LAWS' capable of identifying major weapons platforms such as tanks or submarines as military targets is likely to be relatively feasible, while building those capable of identifying individual humans as either combatant or non-combatant is likely to be very difficult for some time.

The other pertinent Humanitarian Law principle is the *principle of proportionality*, which requires that damage to civilians is proportional to the military aim. Again, current technology for robots can make some defensible judgements on this issue. There is an important exception to this, however, notably in cases where a system could confirm that *no* collateral damage was likely. In practice, judgements of proportionality could be made by the commander and, in the process, specify the domain within which a killer robot would operate. A current example is the process of authorisation of the ship defence system 'Phalanx'.

The relevant commander sets the system to 'weapons free' only in conditions where there is near certainty that fast-flying missile-like objects are hostile and therefore legitimate targets. The same process applies for more potent weapons systems. A descendent technology of the Predator could be given an area of ocean that a commander has good reason to believe will be occupied only by the enemy's navy and therefore could be given 'weapons free' to attack targets within that. The proportionality requirement would thus be met.

The United Nations has recently urged member countries to "ensure that any measures taken or means employed to counter terrorism, including the use of remotely piloted aircraft, comply with their obligations under international law, including the Charter of the United Nations, human rights law and international humanitarian law, in particular the principles of distinction and proportionality" and asserted "the urgent and imperative need to seek agreement among Member States on legal questions pertaining to remotely piloted aircraft operations" (United Nations 2013: 6s, 17). The European Parliament is more forthright and says that "drone strikes outside a declared war by a state on the territory of another state without the consent of the latter or of the UN Security Council constitute a violation of international law and of the territorial integrity and sovereignty of that country" (European Parliament 2014, principle E). Some, notably the U.S. president, have disagreed, and claim wider rights to self-defence (cf. Schmitt 2014).

Given this context, it is unfortunate that the UN Special Rapporteur on extrajudicial, summary, or arbitrary executions, Christof Heyns, reported on LAWS at the same time. The normative issues raised by cross-border targeted killings are one thing; those by the use of LAWS are quite another. In addressing the question whether a killing is extrajudicial, it is irrelevant whether the weapons concerned are used directly, controlled remotely, or function autonomously. In any case, Heyns, who calls for a moratorium, admits, "While it is not clear at present how LARs [lethal autonomous robots] could be capable of satisfying IHL [International Humanitarian Law] and IHRL [International Human Rights Law] requirements in many respects, it is foreseeable that they could comply under certain circumstances, especially if used alongside human soldiers" (Heyns 2013: 109).

This discussion is clouded by a simple dichotomy of autonomous vs. non-autonomous systems. To the contrary, autonomy is a matter of degree and is relational – that is something is autonomous with respect to something else – to some degree (cf. Müller 2012). The vague nature of autonomy has the awkward implication for proponents of a ban that it would be very hard to formulate clear controllable criteria for establishing which systems count as killer robots (cf. Anderson and Waxman 2013).

There is no deep mystery here. The legal situation is clear enough. Weapons that violate the laws of distinction and proportionality in war are already prohibited. Just like any other weapon, killer robots should be used only if doing so is in compliance with the law; otherwise, their use is a war crime. But there is no principled reason to suppose that killer robots could not conform to the requirements of distinction and proportionality. Whether or not they will do so is an empirical question. Contrary to popular belief (e.g. Garcia 2014; Sharkey 2008a, 2008b, 2012), this is *not* an argument to ban killer robots.

It would be an argument only if autonomous killer robots had some feature that would make the prosecution of illegal killings difficult or impossible. Some have suggested that this is the case: "[n]obody is responsible for the killing", they say. This objection deserves more detailed analysis, to which we now turn.

4.2 Responsibility

4.2.1 The Responsibility Gap

The application of criminal law requires that someone be in the dock. This requirement, along with a platitude about the relation between law and morality, generates a serious problem for the justifiability of LAWS. This section sets out the problem. The next summarises how to resolve it.

For the most part, law 'piggybacks' on morality. That is, many legal ascriptions of responsibility do and should track moral responsibilities that are in an important sense prior to the law. For instance, the law should define legal responsibility for murder in ways that ensure the conviction of only those people who have committed the moral wrong of murder.

The problem for killer robots then arises given the following two premises. First, *prima facie*, it will be possible for people to be very seriously wronged as a result of LAWS' action. People who should not be killed – such as civilians, or surrendering soldiers, or prisoners of war – may well be. The moral wrong is of such a degree that there ought to be laws that hold as legally responsible those who are morally responsible for those deaths. A war cannot be waged justly if no one could be morally and legally held accountable for such serious wrongings.

Second, however, it is far from clear who is morally responsible for deaths caused by LAWS. Recall that LAWS are designed to be autonomous, that is to be able to select targets and make 'decisions' to kill people *without* human input. So who is to blame when killer robots get it wrong? It is not clear that the military commander who deployed the system is blame; the commander was just given the equipment along with the order to use it. Nor is it clear that the robot's designers are to blame. They were commissioned to make something that is autonomous; it is a sign precisely of the success of their work that the system is able to do so. Nor, finally, is it the dumb robot. The ability to take on ethical responsibility and be an appropriate target for reward and punishment is a complicated matter that involves at a minimum having goals and desires, as well as the ability to reflect on these, to act against one's desires, and to understand the consequences of one's actions (which is why we usually do not hold small children or animals responsible for their action). It is clear that current systems do not have the necessary properties for responsibility and that, for the foreseeable future, artificial systems will not acquire them – so we do not need a 'machine ethics' in that sense (on this issue, see Gunkel and Bryson 2014).

As Rob Sparrow puts it, killer robots threaten to create a 'responsibility gap'. Yet it is a condition of the possibility of a just war that such gaps not exist (Sparrow 2007; the concept derives from Matthias 2004). This so-called responsibility ascription problem is recognised as a large issue in robotics (on which see

Lichocki et al. 2011; for recent papers on the philosophy of technology, see Battaglia et al. 2014, also Saxon in this volume). As such, killer robots ought not to be used in war. So they ought not to be used at all.

4.2.2 Living with Responsibility Gaps

I deny the thesis that the existence of responsibility gaps morally precludes the use of a technology. The claim is not true in civil life, nor is there reason to think it true in war. Consider the former first. Even in matters of life and death, such as pharmaceuticals or building bridges, we expect only 'due care' from those involved (engineers, public inspectors, users, etc.), and we specify regulations to ensure that the standards of due care are high. But if such care is exercised, we do not require that there be no deaths. A certain engineering 'tolerance' is accepted, even if we try to keep this as small as practically possible. Within that tolerance, we expect the bridge to function correctly, assuming certain environmental conditions. But if these careful estimates were wrong – perhaps we face a stronger earthquake than could be expected – we say that the collapse of the bridge and thus the deaths caused are 'accidental': [n]obody is responsible. Likewise for drugs. There is net benefit in having effective drugs. Drugs are licenced and prescribed, even though we know that there is a risk that some patients will suffer from serious unintended side effects of taking them. This is unavoidable due to the biochemical complexity of humans and the unpredictability of individuals upon the introduction of novel compounds. The benefit of having effective drugs, even with occasional negative side effects, is shared collectively. It is also enjoyed by individuals *ex ante*, who expect in anticipation to gain benefit from taking the drug, even granting the risk of negative side effect.

When a system fails within conditions inside those it should successfully tolerate, someone is responsible. In the bridge case, a driver may be responsible for its collapse if she takes a vehicle over it that is heavier than the load it was designed and stated to be able to bear. Or the engineer may be responsible, for cutting costs by using inferior materials or just being negligent. It could be the public official responsible for confirming compliance with the appropriate regulations. But, importantly, it is conceptually possible that the bridge collapsed and *no one is responsible*. Suppose the bridge crosses a river, which is usually 2 metres in depth and in the last 200 years has never had a volume of water exceeding a depth of 5 metres. But exceptional floods mean that a depth of 8 metres of water is pressing against the pier, which collapses. It would have been a misuse of resources to build the bridge to withstand that force; if the same margins of error were applied to bridge-building generally, it would be inefficient. In such a case, no one is responsible for the bridge's collapse. If this is correct, the real problem is not the possibility that things may go wrong in a way that no one is responsible. Rather, it is how to regulate the tasks of the maker and of the users of the system in such a way that responsibility is *usually* but not *always* ascribable. Further, it is unclear what reason there would be to make special demands for technologies used in war. There is a responsibility gap, but this is the normal gap that we accept in engineering. (This argument is developed at length in Simpson and Müller 2015.) So what I need for my argument here is only the thesis that if a technology produces rare cases of killings where no person is responsible, this does not by itself compel us to ban the use of this technology. A strong responsibility principle that allows no responsibility gaps at all is untenable in practice. However, responsibility gaps are clearly undesirable and would rule out systems that systematically or even just frequently produce such cases. I argue that killer robots can be regulated in such a way that they do not produce more responsibility gaps but rather reduce their frequency: it is already the base that, for many killings in war, responsibility cannot be allocated.

4.2.3 Narrowing the Responsibility Gap

The responsibility framework just outlined shows how responsibility should be ascribed for many of the wrongful killings that could be committed by killer robots. The technology gives rise to a related and further beneficial effect, which is often not noted. Holding someone accountable for their action, such as for actual conviction for a war crime, requires reliable information, which is often unavailable. The ability to acquire and store full digital data records of LAWS' action and pre-mission inputs allows a better determination of the facts and thus of the actual *allocation* of responsibility than is currently possible in the fog of war. As well as allowing allocation of responsibility, the recording of events is also likely to diminish the likelihood of wrongful killings. There is already plenty of evidence that, for example, police officers who have to video their own actions are much less likely to commit crimes. So killer robots would actually reduce rather than widen responsibility gaps (see also Saxon in this volume).

4.2.4 Regulation and Standards

The foregoing has the following implication: moral interest should be focused on the determination of the technical standards of reliability that robots – including killer robots – should meet. The recent EU 'RoboLaw' report makes a parallel point in arguing that we should resist the urge to say that 'robots are special' in terms of responsibility. Rather, we should adopt a functional perspective and see whether the new technology really does require new legal regulation and in which areas (based on Bertolini 2014; Palmerini et al. 2014: 205ff.). This seems to be a move in the right direction: we already devise automated systems (e.g. automated defence of ships against air attacks) where the rules of engagement are put into software. The same due care is to be expected for the manufacture and use of LAWS. Just like for civil autonomous cars, we need to specify standards that LAWS manufacturers must abide by. These standards must ensure that the robot acts according to the principles of distinction and proportionality (this is already possible now if one thinks of targeting tanks, ships, planes, or artillery, for example). Both manufacturing and distributing LAWS that do not abide by these standard would be a war crime. If a killer robot is manufactured with due care according to these standards but commits a war crime, due to use in situations for which it was not designed or licensed, the crime is the responsibility of the soldier/user. The responsible person for a particular command or action can be identified in the military chain of command; this is a deeply entrenched tradition. Finally, if the soldiers can show that they exercised due care, then the deaths are accidents.

The regulation of LAWS thus requires two components. First, there are the technical standards of reliability that LAWS must meet – pertinently, what degrees of reliability must LAWS meet in terms of distinction and proportionality in their decisions to attack? Second, there are the legal instruments by which accountability is to be exercised over those who fail to manufacture, distribute, or deploy LAWS in accordance with those standards. Each dimension – that of technical standards and of the law – should be subject to enforcement at the international and national levels.

The proposed policy structure can thus be schematically presented as shown in the matrix in Table 3.1.

4.3 Are Killer Robots Good for Us?

Assuming there is no particular ethical problem with humanitarian law or responsibility, the remaining question is whether the consequences of LAWS are good overall, in the long run – do they reduce or increase human well-being? If utility is reduced by the existence and use of these weapons, we should not allow them. There are two subordinate questions here. Do LAWS increase or decrease the suffering and loss of war? Do LAWS make war more or less likely?

4.3.1 Reducing the Human Cost of War

A number of points suggest that LAWS would reduce the human cost of war:

- Robots reduce war crimes and crimes in war: they do not rape, they do not get angry or afraid, they do not intentionally commit war crimes – unless programmed to do so. They follow orders more closely and dispassionately.
- "One of the great arguments for armed robots is they can fire second" (Joseph W. Dyer, cited in Markoff 2010).
- Drones are excellent data collectors, so perpetrators of war crimes are more likely to be caught. This also makes war crimes less likely.

Table 3.1 Legal and Technical Regulation

	International	*National*
Legal	International Humanitarian Law	Criminal Law
Technical	Technical standards for performance	Control regimes for technical standards; special standards

- The use of robots will allow for fewer deaths, injuries, and traumas of combatants.
- The use of robots will allow for fewer deaths, injuries, and traumas of non-combatants.
- Therefore, the use of robots will allow for less damage to future generations.

4.3.2 Making War Worse

A couple of points suggest LAWS would make wars worse:

- LAWS have limited judgement and common sense, which will lead to errors and to carrying out orders that violate the law of war.
- Killing is made easier if the command can be passed on to an autonomous system, so proportionality is under threat.

4.3.3 Making War More Likely

Some points suggest that LAWS would make wars more likely:

- With LAWS, a war can be expected to result in fewer deaths and injuries to the soldiers on the side that has them available (but only slightly, if compared to remote-controlled systems).
- They make wars less bad, generally, and thus wars are more likely to be chosen as a means.
- They make a particular military action easier to decide for a military commander (see Krishnan 2009).
- Fewer losses of soldiers' lives reduce the political hurdle for wars and especially military action short of war.
- Finally, robots make it easier to maintain a 'low-level' war for some time, especially if it is an asymmetric war.

4.3.4 Utility, Fairness, and Arms Races

The reasons why LAWS make a war more likely apply equally to remote-controlled weapons; in fact, they apply to any weapon that acts at a distance. Such weapons have always resulted in relative safety for the first users: Goliath probably thought David's sling unfair; the Lateran Council of 1139, Canon 29, banned the use of the large crossbow [*ballista*] and archery against fellow Christians. The criticism that LAWS lower the risk for attackers and thus make wars and other killings more likely is thus *correct* but applies to any weapon that the one side has and the other does not: in other words, it is the result of an ongoing arms race. As soon as the other side has acquired the new weapon, the risk of war goes down again.

This is not to say that I think LAWS are part of an unstoppable arms race. Some weapons systems have been banned (anti-personnel mines, chemical weapons), and with nuclear weapons the arms race is highly controlled. We *can* stop developing these weapons. The question is whether it is right to do so, given that they seem to save lives.

Let us note that the reasons why LAWS make wars less bad do not apply to all weapons at a distance. This is especially so for weapons of mass destruction or weapons with poor accuracy – and thus with poor compliance to the humanitarian law requirements of distinction and proportionality.

If killer robots become cheap and easy to obtain or make, then the consequences would certainly be bad – as in any case of a weapon becoming more widely available (see Müller 2015) – so we would do well to prevent this spread.

4.3.5 Utility Overall

So, what is the overall utility count? As usual with utility in the long run, this is very hard to say, but it seems quite clear that LAWS would do something to reduce the human cost of war. Some reduction is extremely valuable, given how serious the suffering of war is. Further, while LAWS do raise the probability of wars in the short run, there is no reason to think they will do so in the long run. The overall utility calculation depends on the balance of how much less bad wars become and how much more likely they become. How bad the short run rise in probability will turn out depends mainly on who the first parties are to acquire them Given current military spending, we know who these are: the United States, China, Russia. The short-term probability also depends on how big the difference is from current remote-controlled systems, which currently looks minimal. If future systems do not substantially increase the probability of war, then killer robots are good news for humanity.

5. Conclusions and Policy Recommendations

I conclude that:

1 Killer robots pose no substantial challenge to humanitarian law.
2 Killer robots pose no substantial issue of responsibility.
3 Given the first and second conclusions, the crucial issue is whether the overall consequences of killer robots are positive.
4 The overall consequences of killer robots are probably positive.
5 Therefore, we should not ban killer robots.

Given the caveats 1, 2, and 4, I recommend the following policies (for details, see Müller and Simpson 2014):

6 Develop binding technical standards that spell out manufacturers' responsibilities.
7 Maintain a clear chain of command and collect data, assuring responsibility for actions and enhancing provability of war crimes.
8 Affirm and defend the humanitarian law just war requirements, especially clear differences between war and peace, between combatants and civilians

(distinction) and between necessary and unnecessary force (proportionality). These requirements are under threat with current remote-controlled weapons, and this threat will continue with LAWS.
9 Do not ban lethal autonomous weapon systems (LAWS) or killer robots.

Note

1 The policy recommendations of this paper are spelled out in (Müller and Simpson 2014); the philosophical points about responsibility are discussed in detail in (Simpson and Müller Submitted 2015). I am grateful for the cooperation with Tom Simpson. My thanks for comments on earlier drafts go to: Stuart Armstrong, the editors of this volume, an anonymous reviewer, audiences in Delft, Athens, Aarhus and at the online ShanghAI Lectures.

References

Anderson, K, and Waxman, M. 2013. *Law and Ethics for Autonomous Weapon Systems: Why a Ban Won't Work and How the Laws of War Can.* Stanford University, The Hoover Institution. Available online at http://media.hoover.org/sites/default/files/documents/Anderson-Waxman_LawAndEthics_r2_FINAL.pdf.

Arkin, RC. 2009. *Governing Lethal Behavior in Autonomous Robots*. CRC Press.

———, 2013. *The Robot Didn't Do It: A Position Paper for the Workshop on Anticipatory Ethics, Responsibility and Artificial Agents.*Available online at http://www.cc.gatech.edu/ai/robot-lab/online-publications/position_paperv3.pdf.

Battaglia, F, Mukerji, N, and Nida-Rümelin, J (eds). 2014. *Rethinking Responsibility in Science and Technology*. Pisa University Press.

Bertolini, A. 2014. Robots and Liability – Justifying a Change in Perspective, in F Battaglia, N Mukerji, and J Nida-Rümelin (eds.), *Rethinking Responsibility in Science and Technology*. Pisa University Press, 143–166.

Birmingham Policy Commission. 2014. *The Security Impact of Drones: Challenges and Opportunities for the UK* (October). Available online at http://www.birmingham.ac.uk/Documents/research/policycommission/remote-warfare/final-report-october-2014.pdf.

Emmerson, B. 2014. *Report of the Special Rapporteur on the promotion and protection of human rights and fundamental freedoms while countering terrorism.* UN General Assembly, Human Rights Council, 11 March (A/HRC/25/59). Available online at https://documents-dds-ny.un.org/doc/UNDOC/GEN/G12/137/80/PDF/G1213780.pdf?OpenElement.

Enemark, C. 2014. *Armed Drones and the Ethics of War: Military Virtue in a Post-heroic Age*. Routledge.

European Parliament. 2014. *Plenary Session: Joint Motion for a Resolution on the Use of Armed Drones*, 25 February [adopted 27.02.2014] [2014/2567(RSP)].

Garcia, D. 2014. The Case against Killer Robots: Why the United States Should Ban Them. *Foreign Affairs*, 10 May. Available online at http://www.globalpolicyjournal.com/blog/15/08/2014/killer-robots-why-us-should-lead-ban.

Gunkel, D, and Bryson, J (eds). 2014. Special Issue on Machine Morality: The Machine as Moral Agent and Patient. *Philosophy & Technology*, 27: 5–8.

Heyns, C. 2013. *Report of the Special Rapporteur on Extrajudicial, Summary or Arbitrary Executions*. UN General Assembly, Human Rights Council, 23 (3), A/HRC/23/47.

ISS. 2014. International Institute for Strategic Studies: The Military Balance 2014, press statement. Available online at http://www.iiss.org/-/media/Documents/Publications/The Military Balance/MB14 Press Statement.pdf.

Krishnan, A. 2009. *Killer Robots: Legality and Ethicality of Autonomous Weapons*. Ashgate.

Leveringhaus, A, and Giacca, G. 2014. *Robo-Wars: The Regulation of Robotic Weapons*. Oxford Martin School Policy Papers. Available online at http://www.oxfordmartin.ox.ac.uk/downloads/briefings/Robo-Wars.pdf.

Lichocki, P, Kahn, P, and Billard, A. 2011. The Ethical Landscape in Robotics. *IEEE Robotics and Automation Magazine*, 18 (1): 39–50.

Markoff, J. 2010. War Machines: Recruiting Robots for Combat. *The New York Times*, 27 November. Available online at http://www.nytimes.com/2010/11/28/science/28robot.html.

Matthias, A., 2004. The Responsibility Gap. Ascribing Responsibility for the Actions of Learning Automata. *Ethics and Information Technology*, (6):175–183.

Ministry of Defense, Chiefs of Staff. 2011. The UK Approach to Unmanned Aircraft Systems. *Joint Doctrine Note*, 2/11, 30 March.

Müller, VC. 2012. Autonomous Cognitive Systems in Real-world Environments: Less Control, More Flexibility and Better Interaction. *Cognitive Computation*, 4 (3): 212–215.

———. 2015. Gun Control: A European Perspective. *Essays in Philosophy* 16(2): 247–261.

Müller, VC, and Bostrom, N. 2015. Future Progress in Artificial Intelligence: A Survey of Expert Opinion, in VC Müller (ed), *Fundamental Issues of Artificial Intelligence*. New York: Springer, 553–571.

Müller, VC, and Simpson, TW. 2014. *Killer Robots: Regulate, Don't Ban*. University of Oxford, Blavatnik School of Government Policy Memo, November. Available online at http://www.bsg.ox.ac.uk/research/policy-memos/killer-robots-regulate-dont-ban.

Palmerini, E, et al. 2014. *Guidelines on Regulating Robotics*, 22 September. Available online at http://www.robolaw.eu/.

Schmitt, MN. 2014. A Step in the Wrong Direction: The EU Parliament's Drone Resolution. *Just Security*, 7 March. Available online at http://justsecurity.org/7926/step-wrong-direction-eu-parliaments-drone-resolution-schmitt/.

Sharkey, NE. 2008a. Computer Science: The Ethical Frontiers of Robotics. *Science*, 322(5909): 1800–1801.

———. 2008b. Grounds for Discrimination: Autonomous Robot Weapons. *RUSI Defence Systems*, 11 (2): 86–89.

———. (2012), Automating Warfare: Lessons Learned from the Drones. *Journal of Law, Information & Science* 21(2).

Simon-Michel, JH. 2014. *Report of the 2014 Informal Meeting of Experts on Lethal Autonomous Weapons Systems (LAWS)*. High Contracting Parties to the Geneva Convention at the United Nations, 16 April.

Simpson, TW, and Müller, VC. 2015. Just War and Robots' Killings. Available online at http://www.bsg.ox.ac.uk/sites/www.bsg.ox.ac.uk/files/Just_War_and_Robot_Killings.pdf.

Singer, PW. 2009a. Military Robots and the Laws of War. *The New Atlantis*, Winter 2009. Available online at http://www.thenewatlantis.com/publications/military-robots-and-the-laws-of-war.

———. 2009b. *Wired for War: The Robotics Revolution and Conflict in the 21st Century*. Penguin.

Smalley, D. 2014. The Future Is Now: Navy's Autonomous Swarmboats Can Overwhelm Adversaries. *US Office of Naval Research*, 5 October. Available online at https://www.highbeam.com/doc/1G1-386798005.html.

Sparrow, R. 2007. Killer robots. *Journal of Applied Philosophy*, 24 (1): 62–77.

United Nations, General Assembly. 2013. *Resolution: Protection of Human Rights and Fundamental Freedoms While Countering Terrorism*. 68/178. 18 December. Available online at http://www.un.org/en/ga/search/view_doc.asp?symbol=A/RES/68/178.

U.S. Department of Defense. 2012. *Autonomy in Weapon Systems*. Directive 3000.09. http://www.dtic.mil/whs/directives/corres/pdf/300009p.pdf.

———. 2013. *Unmanned Systems Integrated Roadmap* FY2013–2038. Available online at http://www.defense.gov/pubs/DOD-USRM-2013.pdf.

4 Moral Integrity and Remote-Controlled Killing

A Missing Perspective

Bernhard Koch

Responsibility is a multi-term predicate:[1] somebody (*who?*) is responsible for something (*which action? what effects?*) to somebody (*whom?*) because of something (*why?*) at a certain time (*when?*). My considerations focus on one aspect of responsibility: the basic *moral* responsibility for every actor to justify one's actions to oneself. The basis for this justification is normative integrity, which is not completely given through outer factors but in part a result of identity-creating decisions and actions. Being a soldier is different from being a sniper and different from being a henchman – not only in descriptive terms but in normative terms as well. Every person enjoys some liberty for normative self-design, and one's own self-conception plays a major role as a constraining factor against purely consequentialist or objectivist reasoning. Social communities as well can create a normative self-design – and in many ways this is what politics is about. So the question of arming drones for the military is a political one in the first place, but insofar as it is intermingled with the moral self-conception of a community, it is a moral one too.

In short, there seem to be two major advantages of armed drones in modern warfare.[2] (1) The drone pilot can be deployed far away and hence be out of reach of enemy fire. (2) Armed drones seem to spare civilians because of their improved precision. Drones are a fascinating combination of distance and proximity. The protective effect for soldiers is based on their distance; the protective effect for civilians is based on their proximity. Given the protection of soldiers and civilians, it looks as if the usage of armed drones is not only morally permissible but even mandatory. Bradley Jay Strawser has given this important argument a forceful voice:

> I contend that in certain contexts UAV [uninhabited aerial vehicles, or drones] employment is not only ethically permissible, but is, in fact, ethically obligatory. The basis for this claim rests upon what I call the principle of unnecessary risk (PUR). PUR proceeds as follows: If X gives Y an order to accomplish good goal G, then X has an obligation, other things being equal, to choose a means to accomplish G that does not violate the demands of justice, make the world worse, or expose Y to potentially lethal risk unless incurring such risk aids in the accomplishment of G in some way that cannot

> be gained via less risky means. That is, it is wrong to command someone to take on unnecessary potentially lethal risks in an effort to carry out a just action for some good; any potentially lethal risk incurred must be justified by some strong countervailing reason.[3]

Strawser expresses a legitimate perspective. It is the perspective of effects and the perspective of responsibility to (potentially) affected people. I will argue that this perspective is not to be taken as the only one available. There is responsibility to the actor herself as well. If we are willing to accept this, it is not at all clear anymore that effect-based arguments are quite right or cover all aspects that we have to consider morally.[4] My deliberations start from a different point of view, namely the commandment "Thou shalt not kill",[5] but I allow for cases of defensive force in which killing might be justifiable.[6] In this sense, my position is very close to the basic approach of "revisionist just war theory".

Why is Bradley Jay Strawser's perspective – who is a "revisionist" as well – and of most other effect-based proponents of armed drones legitimate? The reason is: better technology is simply better. That's what it is designed for. We improve technology in order to get better results. As soon as better technology is available, it is somehow obligatory to use it. In medical technology, MRI scanners have improved much over the last 20 years. Present MRI scanners are much more precise than 10- or 15-year-old ones. Therefore, it is a legitimate and sensible goal to replace old MRI scanners with new and better ones. If the financial means are available, the replacement is not only permissible but rather compulsory. A former German minister of defence compared the time before the development of armed drones with the age of the stagecoach.[7] Somehow he was right: it would be more than reckless to transport a person with a heart attack in a stagecoach to the nearest hospital. Better technology forces us to use it – just by being better. However, the problem is that "good" and "better" are not simply one-term predicates. Something is good *for* something. Something is better with respect to a purpose or goal.

Drones might be better for villains too. They strive to reach their goals as well and might have better chances to do so with better instruments. You can use instruments for good or bad purposes. Therefore, many people think that, ethically, instruments are indifferent, and most people agree that an unjust war effort cannot be made a just one by using drones. If a party to a conflict is engaged in a military mission without just cause, then every killing is impermissible, even less harmful ones. Jeff McMahan has made the powerful claim that – contrary to international law[8] and a so-called "traditional" version of just war theory represented by Michael Walzer[9] – *ius in bello* is anything but independent from *ius ad bellum*.[10]

But both Michael Walzer's "traditional" account of just war[11] and Jeff McMahan's "revisionist" account of just war come to an agreement that *in bello*, violence is somehow to be justified on the basis of self-defence.[12] Generally, McMahan sees the application of force in wars to be justified on the criteria of *defensive force* – in *ad bellum* terms as well as *in bello* terms.[13] In some cases, lesser-evil justifications are possible. For Michael Walzer, however, combatants are agents of their political community who are not to be held responsible for the good or bad reasons of their community going to war but who are in a situation of mutual self-defence against each other. The difference between Walzer and McMahan in this respect is simply[14] that McMahan distinguishes, with respect to just *ad bellum* causes, between just and unjust combatants, and only just combatants are permitted to fight, whereas Walzer concedes that even combatants whose political community has no good reasons to go to war and who are, therefore, unjust combatants in the respective (McMahan's) sense may fight as well.[15] So we may assume that, at least for just combatants, defence is providing the legitimation to individual soldiers for the use of force in wars. We probably do not have to go into too much detail on the – naturally contested – criteria for defensive force.[16] One very important criterion, however, is that the threat against which one defends oneself or others is somehow 'imminent'. You are certainly not permitted to kill somebody now or today because he could be your murderer next week or next year. Imminence is due to interpretation, and this interpretation is probably the most delicate one. Other very important criteria are necessity and proportionality. Both are two-term predicates; that is for their application you need to direct your reasoning at a goal or an aim. Some action might be necessary for preserving one's life. Some consequences might not be proportional (in a narrow sense) to preserving one single life. Sometimes, it might be possible to escape, so that defensive force is not necessary at all. Sometimes, defensive force might cause enormous harm to bystanders, so that defensive action is not justified as proportionate (in a wider sense).

Let us assume that the revisionist account of just war theory is preferable, and in wars, "just" and "unjust" combatants have to be distinguished. Then, morally speaking, unjust combatants are not permitted to fight at all; just combatants are permitted to fight in self- or other-defence but are restricted to the boundaries of legitimate defensive force taken from a normative conception of self-defence; for example, they are not allowed to make use of bystanders for their own safeguard.[17] The point of interest here is that, as in self-defence, there is a normative asymmetry, and just combatants are not obliged to put themselves at risk.[18] They are victims because they have done nothing to lose any of their basic rights to life and limb. Drones, therefore, seem to be the means of choice. Technology and moral reasoning seem to go together. Taking into account that revisionist just war theory gains more and more popularity,[19] you could say that whoever opposes armed drones and drone warfare can already know today that she or he is on the wrong side of history.[20]

But some people still have mixed feelings about armed drones. What feeds into these? Very often, it is a general scepticism about new technologies and its not yet experienced effects. The first and foremost objection is, therefore, that precisely the previously mentioned advantages in protection could easily turn into disadvantages. Dead or wounded soldiers represent a political burden for politicians in democratic states who go to war too easily. If military missions could be executed with less "political costs",[21] this could make politicians more willing to engage in military actions. In effect, this could even result in more armed conflict and more killed or wounded soldiers. Along a similar line lies the argument that drone operators, being without risk themselves, might be more willing to press the button (the "Playstation mentality"[22]). More killing and more collateral damage could be the result. What is more, drones undermine the protective effects of international law in dissolving the boundaries of restricted battlefields. Drones can be operated from many thousands of miles away, and the operational bases would still be part of the battlefield.[23] Some scholars think that drones are only a further step in military robotics towards total automation ("autonomy") and that there is an unavoidable slippery slope.[24] "Autonomous systems" are, by definition, out of human control. They can be another source of insecurity and danger, as the Campaign to Stop Killer Robots contends.[25] (But Vincent C. Müller contradicts this thesis in Chapter 3 in this volume.)

These and some more arguments offer powerful concerns, but are they able to provide principled grounds to refute armed drones? This is doubtful. The counterarguments are based on conjectures. Some of the assumed effects are probable if nothing is being done to prevent them. Proponents of armed drones could contend that something can be done and that we only need a better normative framework for using drones and an improved accountability regime.[26] In this way, arguments in favour of drones and counterarguments are both based on suppositions about effects, and the ongoing discussion on drones is actually an argument about whether we think that armed drones yield more positive or negative results and how we can get them to yield best results. The so-called ethical discussion boils down to a quarrel about predictions and appropriate countermeasures to prevent bad effects. But ethical thinking is not only about effects. On the other hand, there is a way of reasoning that I would like to label – a bit awkwardly – as "existentialist". This kind of thinking expresses our first and foremost responsibility to justify our action to ourselves. As using armed drones is taking human lives, this thinking is about justifying some actions of killing.

Before I get to my central point, I should mention one aspect that was dominating the discussion on armed drones in Germany. Armed drones were seen as the preferred tool of the U.S. military and secret services for so-called "targeted killings".[27] Chantal Meloni (in Chapter 2 in this volume) presents the history and the practice of targeted killings. I want to focus only on a conceptual point: according to Nils Melzer, targeted killings are defined as having "five cumulative elements"[28]:

1 Use of lethal force
2 Intent, premeditation, and deliberation to kill
3 Targeting of individually selected persons
4 Lack of physical custody
5 Attributability to a subject of international law

The unusual element of this definition for our thinking about war is the targeting of individually selected people. But maybe we can see in it an adaption of the traditional principle of distinction in a way that, in asymmetric conflicts, the name of a combatant becomes something like the uniform in traditional warfare.[29] I shall get back to this point in the next section. The ethically most troubling element is the intentional, premeditated and deliberated act of killing. Even if most combatants in traditional warfare were and are well aware that they must try to kill enemy combatants (and that they themselves could be killed at any time in war), the killing itself nonetheless was not – or should not be – intended as the aim of warfare itself. As the Petersburg Declaration of 1868 has it:

> [T]he only legitimate object which States should endeavour to accomplish during war is to weaken the military forces of the enemy; . . . for this purpose it is sufficient to disable the greatest possible number of men.[30]

So the intentions of combatants fighting combatants in war should be directed only towards disablement – rendering the enemy fighter *hors de combat* – not towards killing.[31] The action of fighting an enemy combatant is successful if the enemy combatant is unable to fight on. But the action of targeted killing is successful – according to its definition – if the enemy combatant is actually dead. This feature brings the action of targeted killing much closer to the execution of capital punishment than to warfare and those who execute it much closer to being hangmen than soldiers. No wonder that politicians and military officials in Germany tried to stress that they have no intentions to use drones for – as they called it – "extra-judicial, illegal killings". But there are two problems with this reassurance. (1) First, the U.S. administration, of course, denies that they are engaged in "illegal killings". On the contrary, they try to show why and how, in their opinion, these killings are justified in legal terms.[32] As it stands, we experience more a fighting about the interpretations of law than law guiding the relevant fighting on the ground. Governments rejecting the practice of targeted killings should state this loudly in order to give their *opinio juris*, which is important when what can be seen as customary international law has to be considered. But a decision made by the German federal attorney (in the case of the targeted killing of Bünyamin Erdogan) gives good reason to think that major elements of the U.S. administration's view are meanwhile accepted even in German *opinio juris*.[33] (2) Second, armed drones are the perfect means for targeted killings. Granted, targeted killings can be executed without drones (e.g. with special forces, as in the putative case of the killing of Osama bin Laden in 2011), and drones can be used for aimless killing. But this is not a very helpful objection. Usually, people buy a technical device in order to achieve the purpose that the technical device has been made for. How plausible is it to assume that people acquire a technical device but constrains themselves to not using the functions that this device is best used for? Who really purchases an iPhone® only for phone calls without using apps? It looks as if drone strikes, when done by drones as appropriately as possible, necessarily collapse into targeted killings or something very similar to it (e.g. so-called signature strikes).

Even if this is correct, targeted killing and the use of drones would not need to be condemned right from the outset.[34] There is a rationale behind the tendency of drone warfare that we could label "the individualisation of war".[35] In antiquity and in medieval natural law theory, political communities are seen as an expression of the social nature of human beings. All human beings were seen as being bound by morality as natural law. In the modern era, the concept of a state and its inner norms seem to be based on a contractarianist view.[36] In this view, all human beings are individuals by nature, but they are bound together by means of a contract in political institutions like a state and we have to distinguish between criminals and enemies.[37] Criminals are domestic law breakers. They live in a political community, but they are violating or have violated certain laws of that community. They do not reject the legal order altogether but make use of it for their ordinary lives. Even the bank robber wants to be protected by ordinary rules of life in a certain community; for example, he wants his stolen goods to be seen as his property and to benefit from the rules that protect one's property in a society. But criminals do not abide by all the laws and violate some of them purposefully. That makes them liable to executive force, which can be seen as twofold: as defensive force (during the act of wrongdoing) and as punishing force (after the act of wrongdoing). In some cases, when no state authority is present, every other member of the community is permitted to exert defensive force – in part on behalf of the political community. Criminals violate the social contract and get punished according to the rules of the contract. Enemies, however, fight against a specific political community and its legal order as such. They usually act not from within the legal and political community but attack it from the outside. In some ways, this deteriorates their status: not being part of the legal community, they can be fought against with lethal force without legal review. In some ways, they enjoy improved status: not breaking any laws (because the laws they are fighting against are not their laws), they are not criminals. Nonetheless, not being protected by the legal system of a certain community, it seems that (almost) everything can be done to enemy fighters. Therefore, the rule of *ius in bello* could only be "[a]ct in a way as to not make further peace impossible".[38] Outrageous actions against one's enemies will surely make peace impossible, but which actions might be seen as outrageous is probably to some extent a contingent fact, to some extent probably not, and the very fact that they are seen as outrageous shows that there is something wrong with these actions.

To think somebody could act against enemies without limitations is against any ordinary moral sense. Our moral intuitions are not contractarianist. Even enemies are human beings, just as we are ourselves. International humanitarian law is an expression of this arbitration between the concept of an enemy and the concept of a fellow human being; it is an arbitration between the traditionalist, collectivist view on war in former times and the individualist idea that every human being should be treated humanely and has even certain pre-political individual rights.

The traditional concept had it that when two political collectives were in a state of war, individual soldiers were only to be seen as – as Rousseau has put it so vividly[39] – pure representatives of their collective, usually their state. The conspicuous sign was the uniform. Every soldier was meant to subordinate his personal integrity to the integrity of the collective. The (non-state) enemies of our time do not fight anymore in uniforms, and very often they do not even fight in the name of a political collective but on their own account – very often religiously inspired. So it isn't implausible – as Michael Gross has suggested – to see the individual name as a proxy for the outdated uniform. (Gross, therefore, speaks rather of "named killing" rather than "targeted killing".) But this means that the individual person gains visibility again, and this social development in warfare is supported by the technological development: not only are individuals visible again, but, with progressively precise weapons systems, they can be attacked more and more individually. Because drones are able to get relatively close to the targets, they seem to be the perfect means for attacking individually selected people. So there is individualisation of warfare from two angles: from those who fight[40] as individuals and from those who are targeted as individuals. And there is growing resistance to the acceptance of collateral damage.[41]

Members of an identical public share, in some ways, a normative relationship. They are aware of normative expectations and hold normative expectations themselves. Globalisation today means a globalised media attention, and this again arguably leads to some – be it rudimentary – normative relationship among all members of the global public. Human rights and International Human Rights Law are the positive legal expression for the tendency in the direction of a global legal community. Even if a fully human-rights-based international order is not our reality yet, we should be able to see this tendency. So, to some extent, a global public and a weak form of a common city of mankind, a "cosmopolis", seem to be forthcoming.[42]

But in a truly cosmopolitan world, "pure" enemies are no longer possible. Terrorists and other fighters are acting "from within", so they are breaking the common law of all mankind. But this challenges traditional modern conceptions of war and *ius in bello*. Wars are becoming more and more law enforcement operations, or "Cosmopolitan Wars".[43] Here again seems to be an embedded argument in favour of armed drones: If wars are actually to be seen as law enforcement operations, why should the law enforcers, the global police, not be permitted to use means that protect the police as much as possible? As there is a normative asymmetry between those who violate the laws and those who enforce them, why could armed drones not be seen as the technical reflection of this asymmetry? Armed drones mirror the normative asymmetry of lawbreaker and law enforcer in their technical asymmetry. Maybe armed drones should be available only to global policemen.[44] But there is a problem: Armed drones can only kill, not capture. Domestic police are, by domestic law and International Human Rights Law, not permitted to resort to lethal force in any and all cases. It cannot be that every lawbreaker in a – even rudimentary – global community simply gets killed. Even for police, killing is permissible only when they act in self-defence. And here seems to be the rub.[45]

If indeed the legitimacy of applying military force is ultimately to be found in considerations of defence, I think we come closer to an exploration of what feeds into the feelings of those people who reject armed drones. At first, it is probably helpful to distinguish two ways, or perspectives, to look at armed drones.[46] One perspective is the perspective of a user. In our case, these are soldiers, officers, generals, and, to some extent, politicians: in short, people engaged in warfare. The other perspective is the perspective of theorists. Users look at drones and ask what they are in practice. Theorists look at drones and ask what they are in principle. I think theorists should not devaluate the perspective of users. But on the other hand, practitioners should be open-minded about the point of view of theorists because, when it comes to responsibly justifying one's actions, every user takes on the point of view of a theorist – at least to some extent. To say that a perspective is wrong means to have a superior perspective from which you could judge the rightness or wrongness of subordinate perspectives.

To distinguish between different perspectives means to realise that there are not only one's own arguments in favour of or against armed drones but that one's own look at the issue does not give the whole picture. The missing perspective I would like to bring into these reflections is twofold: One aspect is based on the fact that using drones despite their surgical precision is still the application of (lethal) violence.[47] The other aspect is that drones are designed to withdraw users from any risk or danger to themselves. Armed drones as remote-controlled weapon systems are – in a very principled perspective – the *total* instrument. Users are not only 1 metre, 100 metres, a mile, or 5,000 miles away. In principle, they are *nowhere.* They have completely withdrawn.[48] This is the difference between a bow, an arrow, or artillery and armed drones. The idea of a remote-controlled system like a drone is to put *all* the burden of instrumental use on the instrument and to give *total* relief to the user. This idealistic perspective made the philosopher Walter Benjamin contrast drones with suicide bombers as early as 1935–1936.[49] Both are forms of a total usage of technology. But whereas the suicide bomber uses himself as an instrument and is a total instrument in this sense, the drone operator withdraws completely and makes her instrument a total instrument.

The crucial problem with drones is that drones are a means to apply force in (necessary) defence, but it's exactly the means that tries to prevent a situation of necessary defence. A common objection to this point is that cruise missiles are a distancing means in warfare as well. But drones as remote-controlled systems are *in principle* meant to bring the actor of violence out of reach of the victim of violence. As the difference in practice seems to be marginal, the difference in conceptual terms is often overlooked. To distance oneself *as far as possible* is not the same thing as to distance oneself *completely.*

It might well be possible that thinking about drones and their legitimate usage might force us to reconsider the legitimacy conditions of the usage of cruise missiles. Cruise missiles are launched from a great distance, but nonetheless they are not launched from nowhere, whereas drones are meant to be launched from nowhere – at least from the enemy's nowhere. So, practically, the usage of drones and the usage of cruise missiles might seem to be similar, but from the conceptual point of view, they are not.

This gets us to the problem with defence. In the ideal drone scenario, the application of violence is caused by an actor who is not exposed to any threat. In this ideal drone scenario, the argument of defence cannot hold any more. Conceded: some present usages of armed drones may be based on situations of other- or third-party-defence, for example in providing close air support for convoys.[50] Drones may be used in this way, but present-day models of MALE drones are made not for scenarios like this but rather for targeted-killing scenarios in which no person is under *immediate* threat.[51]

But be it as it may, I want the reader to think only about the conceptual premises in drone warfare, and my thesis is that armed drones, by reason of their being remote controlled, undermine the legitimatory basis of their application in the use of force.

Some defenders of armed drones might object that I could be demanding some code of chivalry that compels well equipped soldiers to throw their arms away in order to be exposed to more or greater risk. But this is a misrepresentation. It is a gripping question to ask what might have motivated a chivalrous fighter in former times to drop his own sword when he saw that his adversary's sword had been broken. But this question is beside the point here. *I do not want to say that soldiers should give up on their protection, but I want to say that their standard of protection might make it impermissible to use violence against their enemies.* The reason is that the argument of self-defence (and in most cases the argument of other-defence as well) falls flat. This is the principled problem with drones. Bradley Jay Strawser's argument, which I have previously quoted, holds, as he himself says, "other things being equal". But with drones, in most cases, other things are not equal; the threat has changed or even diminished. *It is not (the exposure to) risk that legitimizes force towards another person* – and for my argument it is irrelevant whether drone operators are still under certain risks, as Jesse Kirkpatrick in the following chapter explains[52] – *but the absence of risk or (better) a threat that delegitimizes force towards another person.* Or to put it another way: *I don't want soldiers to give up their protection, but if fully protected, I want them to give up on killing.*

In part, Strawser's and others' kind of deliberation is a result of thinking about legitimate force in terms of individual liability. So the first step of one's reasoning is whether A is liable to be attacked. If so, I am justified in attacking that individual. The second step is which are the best, that is least risky, means for the attack? I do think this kind of double-step reasoning bears some kind of moral gravity, but, employing it as a generalized tool, the user of military force might lose the perception that A is not liable anymore because, with remote-controlled devices, the threat is gone.[53] Of course, as long as drones are used in situations of other-defense, for example as a protection for convoys, there might be legitimate ways to do so. Unfortunately, I cannot go into discussing the issue whether the conditions for legitimate self- and other-defense are the same.[54] But just as it would be hard to understand why persons should give up standards of protection for themselves, it might be very hard to understand why others should still be exposed to enemy threat when means of warfare like drones are available. A pure logic of protection forces us to draw back as much as possible and make as much use of drones as we can. But then our enemies can do no more than attack our "material property" (the drones), but this could probably not justify killing them. But that's exactly what armed drones are made for. There is a paradox.[55]

This paradox can be well stated from the perspective of an independent third-person point of view, and this is what I have tried to do so far. But ethical deliberation is not finished with looking at an issue from a third-person perspective. I think it requires more: a first-person perspective, because there is a basic responsibility for every actor to justify one's actions to oneself. Therefore, I would like to have a look at the moral perspective of drone operators. They are agents of violence. Their acting kills people, and they are aware of this. But they are, as well, aware that they are not under threat themselves. So it is not an act of self-defence. In this respect, their acting resembles more the acting of hangmen than that of soldiers. But it might be an act of third-party defence, such as protection of fellow soldiers or civilians. This, by the way, does not invalidate the comparison with hangmen because even hangmen can argue that their action is meant to protect society – other people – from a threat.[56] This similarity again need not invalidate that there might be a good reason for the drone pilot to act in the relevant way. Among the existential problems might be the one that the drone pilot will be aware that she is not acting *as a soldier*. Being exposed to no risk imposed by adversaries[57] is somehow incompatible with the professional normative role of soldiers. Again, this does not mean that one's acting is simply impermissible, but it does mean that it is not a way of acting within some kind of professional normative integrity as a soldier.[58] In some way, it comes closer to the professional normative integrity of executioners, conceded that hangmen are not preventing people from imminent threats, which could be said of drone pilots (or the integrity of snipers who act only in situations of very imminent threat). But without imminent threat in acts of targeted killings, the acting of drone pilots is very similar to that of hangmen: bringing people to death as an executioner of a state order.

Let me explore this existentialist perspective with two examples in the literature on war and killing: (1) Killing without being in danger oneself is taking a life without being able to say, "It's either you or me." This is one of the conditions that Thomas Nagel found important in his famous but scarcely understood paper on "War and Massacre":[59]

> Whatever one does to another person intentionally must be aimed at him as a subject, with the intention that he receive it as a subject. It should manifest an attitude to him rather than just to the situation, and he should be able to recognize it and identify himself as its object.[60]

Nagel takes the violence and the harm that are done to an adversary very seriously. Are his conditions fulfilled in drone strikes? Can a remote-controlled robotic machine really express an attitude towards somebody as a subject? Can the pilot do the same – miles or thousands of miles away? (2) In "L'homme revolté" ("The Rebel"), Albert Camus sides with the 18th-century philosopher Marquis de Sade (1740–1814) in the rejection of capital punishment.[61] The reason is that killing may never be done in a purely impersonal, representative way, as it is done by state executioners.[62] Violence might seem to be unavoidable in certain circumstances, but it has to be the expression of an interpersonal relationship, for example in a conflict about non-shared values.

A thoughtful consideration of this missing perspective might even feed back into a functionalist perspective as well because the omission of the fact that violence has to be based on personal attitudes might be part of the reason why Western military missions[63] in countries with a different culture seem to be so untrustworthy. With drones, it might actually be impossible "to win hearts and minds".[64] This is so because those who apply drones expose their hearts as people solely concerned about their perfect protection but not concerned about the values they pretend to stand for. The way somebody fights affects and is affected by the goal she is fighting for. It seems to me that we can find a deeper reason here why drone strikes have turned out to be a recruitment assistance in the relevant countries for – often young – people to become suicide bombers. In a report of the Stimson Center from 2014, the members of a task force led by General John P. Abizaid and Rosa Brooks contend, "Even strikes that kill only terrorist operatives can cause great resentment, particularly in contexts in which terrorist recruiting efforts rely on tribal loyalties or on an economically desperate population".[65] The authors think this is "in part due to the perception that such strikes cause excessive civilian deaths, and in part due to concerns about sovereignty, transparency, accountability and other human rights and rule of law issues". It's doubtful whether people in these countries really worry about sovereignty or human rights – which they may do. It seems more plausible that they simply think that drone strikes are an expression of Western cowardice to fight for their (Western) values.

Armed drones communicate Western values but not the ones Western states have in mind when they engage in military missions elsewhere: freedom, law, human rights. They rather communicate pure and crude technical power and full protection for actors of violence. In this way, drones are unable to contribute to peace in a manner that affects "hearts and minds", that is to bring about a proper peace. All you can achieve with drones is technical control and the conditioning of human behaviour. This works well as long as humans are seen only from the technical side of their existence. Drones can help to steer and control human behaviour. But an anthropology that focuses only on behaviour is far too narrow. To approach the nature of human beings without reference to the importance of interpersonal relations is insufficient. Human beings are more than only technical beings. Human beings are able to have normative opinions, convictions, values and a sense for human life. Peace is not just an issue of behaviour. It is also an issue of convictions and values. Even if drones are successful in the steering of behaviour, they will nonetheless be unable to bring about peace.

Western functionalist thinking seems to force us to use the most protective technology available. Hereby it is oblivious to the fact that protection is only one orientation on a normative self-conception, which can be balanced against other orientations – just as (physical) pleasure can be balanced against orientations of a greater moral significance. An international lawyer of the University of Bonn, Stefan Talmon, claimed that, according to human rights law, soldiers can make a human-rights-based claim to get the best protective equipment available, among them armed drones.[66] Put aside that armament is probably not the most important feature with respect to protection but is rather needed for fighting one's enemies, and let us assume that Talmon's contention is correct and that soldiers could sue their state at the European Court of Human Rights: it is still by no means inevitable to put all the weight on protection. We can *responsibly* (!) opt for other self-conceptions as well – without saying that protection is futile. Imagine the following case. There are two couples. Both live quite a distance from any other human settlement. Both have the same age and the same income, and both need a new car. Couple A opts for a BMW X4®. Strong, fast, safe, reliable – that's what the car should be. Couple B opts for a Smart® because they put special emphasis on climate and environment. After both couples have gotten their respective vehicles, both men get sick with a very similar heart attack. Lady A is able to get her partner to the hospital in time. Lady B arrives too late. Has couple B been acting in a *morally irresponsible* way? Could the children of B sue their mother in court for not using the best protection available for their father? Is there a duty to have a BMW X4? Those who think that this conclusion of a duty to have the BMW is simply absurd might agree that human beings are permitted to decide on the relationship of protection and other values that they find important. They might agree that people can *responsibly* give up some protection. Morality is not just a matter of deductive reasoning. It is also a matter of constraining oneself and giving oneself normative roles.[67] Sometimes, people are forbidden to do certain things not because of overall objective and deductive reasons but because of their own restrictions. Someone might reasonably say that eating meat is worse when it is done by a vegetarian because in being a vegetarian there is a kind of moral self-determination. A vegetarian eating meat undermines his own moral self-determination and commitment, notwithstanding that even non-vegetarians have very good reasons to relinquish meat. A physician, on the other hand, has probably more duties in helping people than non-physicians. In part, this is so because, as a physician, she received special training in helping people with physical problems. But in part this is also because she gave herself a self-determination as somebody devoted to helping people.[68]

The same holds for political communities. It is not simply unreasonable to opt against armed drones because of certain values that are highly esteemed in a society. It is a matter of political arbitration and decision-making. Political communities *are responsible* for their moral self-conception as well.

Armed drones can easily give you the elusive impression that security for soldiers can be improved *ad infinitum*. If this were true, the very first consequence would be that the profession of soldiering dissolves. The outline of a profession is not only given in descriptive terms but in normative terms as well. Soldiers are risk bearers. The wish to reduce the risks is comprehensible and not illegitimate. But there is a tendency in it to dissolve the profession of soldiers as such – something that in itself has a comprehensible and not per se illegitimate side. Many pacifists might rejoice. Being a soldier means to accept risks, put enemies under risk, and prevent other people (civilians) from certain risks. Soldiers are risk takers. Drone pilots are not. As long as there are soldiers, they have to bear risks, and this means that there cannot be total protection for them. I think Michael Walzer is right in saying that we can expect soldiers to bear risks for civilians – but not only for civilians.[69] Every norm of *ius in bello* can mean that soldiers have to face some disadvantages – up to risking their own lives. Take the protection of cultural heritage. It is easy to imagine situations where soldiers might be better protected if this or that building was not to be spared by international agreements. Sometimes, the protection of cultural heritage can lead to more risks for soldiers and, in the end – tragically – to some additional dead soldiers. But should we do away with all the agreements on cultural heritage protection? Risk depends on preferences.[70] Somebody climbing a high mountain is facing some risk of not surviving this present day. Someone who tries to commit suicide is facing some risk to survive. Some wars are fought about the right preferences.

What we conceive as risk is dependent on our value conceptions. 'Fighting' with the help of drones is an expression of the importance we give to securing our individual lives. This is not irrational. It is the very way of our responsibly thinking that forces us to improve security for human beings wherever we can. But what for? What kind of world are we about to bring about? What is our eschatology? What is our perspective on time and future? Barack Obama was right when he said at the National Defense University:

> From our use of drones to the detention of terrorist suspects, the decisions that we are making now will define the type of nation – and world – that we leave to our children.[71]

I assume adversaries of the Western military who are fought with drones do have answers to these questions. They probably have answers that we would not care to hear. But do we have answers? It seems that the risk aversion of Western soldiers and their claims to get maximum protection attest that they actually do not believe that their cause is worthwhile.

Notes

1 Ropohl (1993, 155). Cf. Lutz-Bachmann (2008, 75); Werner (2013).
2 Cf. Introduction to this volume.
3 Strawser (2010, 344). Uwe Steinhoff challenges Strawser's argument in Steinhoff (2013, 197ff.). On drone warfare in general, cf. e.g. Springer (2013). On the ethical debate on drone warfare, cf. Strawser (ed.) (2013),
4 By "perspective", I mean a way of looking at things taken from a standpoint and having a blind spot in looking at the object. To find out about one's blind spots, you have to change the perspective.
5 Which is more than only a "biblical" commandment (Ex, 20,13; Deut. 5,17) and deeply embedded in our ordinary moral thinking. I agree with Jeremy Waldron here that "the default position, apart from any convention, is that intentionally killing or attacking any human being is prohibited as murder" (Waldron 2010, 106).
6 The effect of killing might be justifiable, not the action of killing. Cf. note 31.
7 Dr. Thomas de Maizière, in a speech delivered at the German Bundestag 31 January 2013.
8 Cf. e.g. Preamble of Protocol Additional to the Geneva Conventions of 12 August 1949, relating to the Protection of Victims of International Armed Conflicts (Protocol I), 8 June 1977.

9 Walzer (2006).
10 McMahan (2004, 2009).
11 Jesse Kirkpatrick (in this volume), like some others, calls it "orthodox approach" (page 110 in this volume). Alex Leveringhaus (in this volume) refers to this debate in Fn. 2.
12 And so do many others in the "revisionist camp", e.g. Frowe (2014).
13 Pacifist accounts traditionally reject this application of legitimate self-defense as a justification for war. Cf. Ryan (2013).
14 This remark is not meant to level the differences in the conceptions of defensive force. McMahan has meticulously elaborated his impressive account.
15 That does not mean that Walzer's account is necessarily more permissive. For example, for Walzer, civilians are not part of the war game and must not be attacked directly. For McMahan, direct attacks on civilians could be justified on defensive grounds but only in very rare real cases.
16 A good overview on the current concepts is given in Robinson, Garvey and Ferzan (eds) (2009), esp. chap. 18, 385–406. McMahan calls his approach the "moral responsibility account of defensive force". Liability depends, in the first place, on moral responsibility for a threat. "Culpability accounts" are stricter: responsibility is not enough. Liability depends on the culpable causation of a threat.
17 On the Lockean account. "According to the Hobbesian ethics of self-defense, you violate no duty if you shield yourself from the bullets that are about to hit you by using a bystander who just happened to be around (in case this is necessary for your survival)" (Benbaji 2008, 465).
18 According to Michael Walzer's account, combatants – there is no distinction of just and unjust ones – have to take on risks in order to spare civilians (because war is the business of combatants), but they do not have to accept risks in order to spare enemy combatants.
19 For many good reasons. For example, it sounds pretty implausible that in humanitarian interventions, the moral status of the just intervening forces should be considered as morally equal to the defenders of the rights-violating regime.
20 As Van Wynsberghe (in this volume) shows, we will get more and more civilian uses of drones as well.
21 Müller/Schörnig (2010, 18).
22 Dobbing, Hailwood, and Cole (2010).
23 This argument is especially interesting with respect to U.S. drone bases in Germany: could Germany be a target if drones were operated from within this country? In legal terms, it could.
24 Dickow and Linnenkamp (2012).
25 www.stopkillerrobots.org Cf. Human Rights Watch (2012, 2014). But Human Rights Watch is focussed only on conjectured effects. There would be a much deeper sense of losing humanity with fully automated systems that has to do with the way we encounter these machines and how we understand ourselves confronted with technology that is designed to outsource violence.
26 This is the line of argument given by Buchanan/Keohane (2015).
27 This practice is seen as a violation of basic human rights because it seems to be the exercise of capital punishment without due process or judicial review. Cf. Heinz (2014).
28 Melzer (2008, 3ff.).
29 This proposal has been made by Gross (2006).
30 Available online at http://www.icrc.org/ihl/INTRO/130?OpenDocument (3 July 2015).
31 This important aspect was recently stressed independently by Nigel Biggar and Robert G. Kennedy (Biggar 2013, 92–110; Kennedy 2014). This line of thinking goes back to Thomas Aquinas (S. th. IIa IIae 64, 7).
32 The justifications started with Harold H. Koh's speech at the Annual Meeting of the American Society of International Law (ASIL) on 25 March 2010 on "The Obama Administration and International Law". Available online at http://www.state.gov/s/l/releases/remarks/139119.htm (3 July 2015).

33 Available online at https://www.generalbundesanwalt.de/docs/drohneneinsatz_vom_04oktober2010_mir_ali_pakistan.pdf (3 July 2015).
34 On targeted killing, cf. Koch (2015, 2011).
35 Cf. e.g. Blum (2013).
36 "The moral theory of contractarianism claims that moral norms derive their normative force from the idea of contract or mutual agreement" [Ann Cudd, "Contractarianism", *The Stanford Encyclopedia of Philosophy* (Winter 2013 Edition), Edward N. Zalta (ed.)]. Available online at http://plato.stanford.edu/archives/win2013/entries/contractarianism/.
37 Cf. Koch (2014a).
38 Cf. Immanuel Kant's sixth "Preliminary Article" in Kant (2006, 70ff.).
39 Cf. Rousseau (1997, 46ff.).
40 E.g., those who believe in a postmortal reward for their fighting fight in their individual interest.
41 Cf. the concept of *ius ad vim*: Brunstetter and Braun (2013).
42 Cf. Marshall McLuhan's talking about a "global village" [McLuhan 1994, 89–105 (orig. 1964)].
43 Cf. Fabre (2012), who follows Jeff McMahan in many ways under the headline of cosmopolitanism.
44 In a German newspaper commentary, the commentator made the point that the German Bundeswehr should use armed drones because even the UN is using them (Müller 2014). But the conclusion might be wrong: maybe *only* the UN is permitted to use drones because the UN has enough authority as a global police agency.
45 I do not want to say that the problem is technical asymmetry as it is discussed in Enemark (2014, chap. 5, Radical Asymmetry). In terms of technical superiority, war is almost never symmetrical, and even "chess" is – contrary to Enemark's assertion (p. 73) – not "the perfectly symmetrical game" as every chess player knows. There is an advantage for the player who is permitted to make the first move.
46 I am talking only about so-called MALE-UAVs (medium altitude, long endurance, unmanned aerial vehicles). It is enthralling and depressing to think about very small, insect-like armed drones, which might be available in some decades.
47 Some might say, the application of force. But this is why the German discussion is interesting. There is just one word for violence and force: *Gewalt*.
48 Cf. Koch (2014).
49 In earlier versions of his famous essay on "The Work of Art in the Age of Mechanical Reproduction", now finally edited in Benjamin (2013). Benjamin contrasts what he calls "first technology", which uses humans as much as possible, and "second technology", which uses humans as little as possible. In his own French version of the text, he says: "La première engageant l'homme autant que possible, la seconde le moins possible. L'exploit de la première, si l'on ose dire, est le sacrifice humain, celui de la seconde s'annoncerait dans l'avion sans pilote dirigé à distance par ondes hertziennes" (173). For German versions, see pages 108 and 150. Cf. Chamayou (2013, chap. II).
50 This argument is prevalent in German discussions about arming German Armed Forces (Bundeswehr) with armed drones. Cf. Speech of the German Minister of Defence, Dr. Ursula von der Leyen, at the German Bundestag 2 July 2014. Available online at https://www.bundestag.de/dokumente/textarchiv/2014/kw27_de_drohnen/286034 (3 July 2015).
51 There may be scenarios in which the use of drones might be justified. I think of an IS fighter who is about to decapitate a hostage with his knife, an action that is widely – actually almost unanimously – considered to be unjustified. There seems to be an obvious permission to kill the IS fighter. If we assume that the defensive killing could be done with a gun or with a drone (*ceteris paribus*), the argument that the drone provides better protection for the defensive killer is very forceful. My intention here is not to say that drones are bad in themselves but that we have to be aware of the different facets of this very radical means of killing.

52 Jesse Kirkpatrick is certainly right that even drone operators are under certain risks but these risks are different from the risks that soldiers face. *Soldiers face risks that adversaries or enemy combatants impose on them. Drone operators face risks that are connected with their own acting.* In this sense almost all professions involve risks (e.g. of burnout). The situation of drone operators is in this sense closer to the situation of casualty doctors. The victim of an accident usually is not responsible for the trauma of the physician.

53 Even in international law, there are similar considerations. Justifications of the use of drones rely heavily on a permission given in the Law of Armed Conflict. But the Law of Armed Conflict is applicable only when no effective control prevails. As soon as the effective control condition is fulfilled, human rights laws apply. But drones could be seen as a means of gaining effective control – so effective that armed drones rule out their own application. On the conditions of liability in McMahan's account, cf. McMahan (2011).

54 Basically German legal interpretation holds that there are no special requirements in cases of other-defence (Lackner and Kühl 2014, 269). But in a moral perspective, there may be reasonable doubts. Cf. Cécile Fabre's "personal prerogative justification for self-defence" (Fabre 2012, 57), which holds that an innocent victim may kill an innocent person in an act of defense because that person does not mean to pose a threat and therefore give greater weight to the preservation of her own life. Such a position may have consequences with respect to the permissible "amount" of force as well.

55 Christian Enemark comes close to my point when he compares this paradox with Joseph Heller's Catch-22: "Removing oneself from risk completely is the most rational thing a person could want, and yet only by experiencing at least a scintilla of risk can a killer claim to be engaged in a warlike activity invested with moral potential" (Enemark 2014, 96). But I would like to substitute "warlike" with "violent". Moreover, the paradox here is similar to Paul W. Kahn's "The Paradox of Riskless Warfare" (Kahn 2002, 2–8). Kahn sets out with a *normative* conception of war and a *normative* conception of policing and contends that modern warfare resembles more policing than war. But then it is inconsistent to apply to justifications for violent force that derive from the war paradigm as, for example the permission to kill enemy combatants. I agree with his conclusions in many ways.

56 A executioner might probably argue that he does not act himself but is just an instrument of the state's acting.

57 Cf. note 46 on Jesse Kirkpatrick's contention that drone operators are facing risks.

58 I use the term "integrity" in Bernard William's sense. The acknowledgement of moral integrity is the appreciation of "the extent to which *his* [an actor's] actions and *his* decisions have to be seen as the actions and decisions which flow from the projects and attitudes with which he is most closely identified" (Williams 1973, 116ff.).

59 Nagel (1972, 137).

60 Nagel (1972, 136).

61 Camus says de Sade "was never able to tolerate legal crime" and quotes de Sade as saying, "To kill a man in a paroxysm of passion is understandable. To have him killed by someone else after serious meditation and on the pretext of a duty honorably discharged is incomprehensible" (Camus 2013, 18).

62 A very impressive account of this ethos is given by the German Catholic author Gertrud von LeFort in her famous novel, *The Song at the Scaffold*, about the killing of Carmelite monks during the French Revolution: "But do not think, my friend, that at this point I expect you to visualize the bloody guillotine! I myself cannot endure the sight of that horrible machine. Believe me, I had rather see a living executioner at work, a man who has the courage to wield the knife, and a hand of flesh and blood that knows at least that it is perpetrating an awful deed. Life should not be shattered by machinery" (trans. by Olga Marx, San Francisco, 2011: 99ff.).

63 On the Western Way of War as "risk-transfer war", cf. Shaw (2005).

64 It may be that the idea of winning hearts and minds is itself a pretty "Western" idea because it requires the distinction of an outer person and an inner one. Sometimes it seems to me that in different cultures, this distinction is of different significance.
65 (The Stimson Center 2014, 10/11).
66 Talmon (2014).
67 In this sense, my considerations are indebted to Cicero's presentation of the four "personae" that we play as moral actors (off. 1, 107–115). The forth role "which we should like to play is prompted by our own choice. So some devote themselves to philosophy, others to civil law, and others again to eloquence; and even in the practice of the virtues different people prefer different ones at which to excel" (Cicero 2000, 39ff.).
68 She is a "doctor", not a "schmoctor" in Arthur I. Applbaum's sense (Applbaum 1999, 45–60). A Christian has even better reasons than non-Christians to abstain from violence. Cf. Hauerwas (1985). Hauerwas's "kind of pacifism forces us to consider the kind of persons we ought to be so that certain kinds of decisions are simply excluded from our lives" (101).
69 Walzer (2004).
70 "Risks involve threats to outcomes that we value. Defining risk means specifying those valued outcomes clearly enough to make choices about them" (Fischhhoff and Kadvany 2011, 22).
71 On 23 May 2013. Available online at http://www.whitehouse.gov/the-press-office/2013/05/23/remarks-president-national-defense-university (3 July 2015).

References

Applbaum, AI. 1999. *Ethics for Adversaries. The Morality of Roles in Public and Professional Life*. Princeton: NJ: Princeton University Press.

Benbaji, Y. 2008. A Defense of the Traditional War Convention. *Ethics* 118: 464–495.

Benjamin, W. 2013. Das Kunstwerk im Zeitalter seiner technischen Reproduzierbarkeit, in Burkhardt Lindner (ed). *Walter Benjamin Werke und Nachlaß. Kritische Gesamtausgabe*, vol. 16. Berlin: Suhrkamp.

Biggar, N. 2013. *In Defense of War*. Oxford: Oxford University Press.

Blum, G. 2013. The Individualization of War. From War to Policing in the Regulation of Armed Conflict, in Austin Sarat, Lawrence Douglas &Martha Merrill Umphrey (eds). *Law and War*. Stanford, CA: Stanford University Press, 48–83.

Brunstetter, D, and Braun, M. 2013. From *Jus ad Bellum* to *Jus ad Vim*. Recalibrating Our Understanding of the Moral Use of Force. *Ethics & International Affairs* 27(1): 87–106.

Buchanan, A, and Keohane, RO. 2015. Toward a Drone Accountability Regime. *Ethics & International Affairs* 29(1): 15–37.

Camus, A. 2013. *The Rebel*. London, Penguin Classics.

Chamayou, G. 2013. *Théorie du drone*. Paris: La Fabrique Editions.

Cicero, MT. 2000. *On Obligations*. Translated with an Introduction and Notes by PG Walsh. Oxford: Oxford University Press.

Dickow, M, and Linnenkamp, H. 2012. *Kampfdrohnen – Killing Drones. Ein Plädoyer gegen die fliegenden Automaten, SWP-Aktuell 75*. SWP_Aktuell. December. Available online at http://www.swp-berlin.org/fileadmin/contents/products/aktuell/2012A75_dkw_lnk.pdf (3 July 2015).

Dobbing M, Hailwood A, and Cole C. 2010. *Convenient Killing. Armed Drones and the 'Playstation' Mentality. The Fellowship of Reconciliation*. Oxford: Oxford University Press.

Enemark, C. 2014. *Armed Drones and the Ethics of War. Military Virtue in a Post-Heroic Age*. Abingdon: Routledge.

Fabre, C. 2012. *Cosmopolitan War*. Oxford: Oxford University Press.
Fischhhoff, B, and Kadvany, J. 2011. *Risk. A Very Short Introduction*. Oxford: Oxford University Press.
Frowe, H. 2014. *Defensive Killing*. Oxford: Oxford University Press.
Gross, ML. 2006. Assassination and Targeted Killing. Law Enforcement, Execution or Self-Defence. *Journal of Applied Philosophy* 23(3): 323–335.
Hauerwas, S. 1985. Pacifism. Some Philosophical Considerations. *Faith and Philosophy* 2(2): 99–104.
Heinz, WS. 2014. *Wann hat der Staat das Recht zu töten? Gezielte Tötungen und der Schutz der Menschenrechte*. Deutsches Institut für Menschenrechte Policy Paper No. 23. Berlin: German Institute for Human Rights.
Human Rights Watch. *Losing Humanity. The Case against Killer Robots*, November. Available online at http://www.hrw.org/reports/2012/11/19/losing-humanity-0 (3 July 2015).
Human Rights Watch. *Shaking the Foundations. The Human Rights Implications of Killer Robots*, May. Available online at http://www.hrw.org/reports/2014/05/12/shaking-foundations (3 July 2015).
ICRC. 1977. *Protocol Additional to the Geneva Conventions of 12 August 1949 and Relating to the Protection of Victims of International Armed Conflicts (Protocol I), 8 June 1977*. Available online at https://www.icrc.org/ihl/INTRO/470.
Kahn, PW. 2002. The Paradox of Riskless Warfare. *Philosophy & Public Policy Quarterly* 22(3): 2–8.
Kant, I. 2006. *Toward Perpetual Peace and Other Writings on Politics, Peace, and History*. Edited by Pauline Kleingeld, translated by David L. Colclasure. New Haven and London: Yale University Press.
Kennedy, RG. 2014. Dürfen Soldaten vorsätzlich töten? Eine theologische Untersuchung, in *Den Gegner schützen? Zu einer aktuellen Kontroverse in der Ethik des bewaffneten Konflikts*. Edited by Bernhard Koch. Baden-Baden, 241–263.
Koch, B. 2011. Ein Beigeschmack von Selbstjustiz. Lässt sich das gezielte Töten von Terroristen rechtfertigen? *Herder Korrespondenz* 65: 352–356.
Koch, B. 2014a. Zur Debatte um den Einsatz von Drohnen. *Stimmen der Zeit* 11: 779–782.
Koch, B. 2014b. Zum Verhältnis von Freund und Feind im bewaffneten Konflikt. *Informationes Theologiae Europae* 18: 223–239.
Koch, B. 2015. Targeted Killing. Grundzüge der moralphilosophischen Debatte in der Gegenwart, in Veronika Bock, Johannes Frühbauer, Arnd Küppers, and Cornelius Sturm (eds), *Christliche Friedensethik vor den Herausforderungen des 21 Jahrhunderts*. Baden-Baden, 191–206.
Lackner, K, and Kühl, K. 2014. *Strafgesetzbuch. Kommentar*. München 28.
Lutz-Bachmann, M. 2008. Kosmopolitische Verantwortung. Über Ethik und Recht in einer globalisierten Welt, in Oliver Decker and Tobias Grave (eds), *Kritische Theorie zur Zeit*. Leipzig: Springer, 70–77.
McLuhan, M. 1994. *Understanding Media. The Extensions of Man*, Cambridge, MA, 89–105 (original 1964).
McMahan, J. 2004. The Ethics of Killing in War. *Ethics* 114: 693–733.
McMahan, J. 2009. *Killing in War*. Oxford: Oxford University Press.
McMahan, J. 2011. Who is Morally Liable to Be Killed in War? *Analysis* 71(3): 544–559.
Melzer, N. 2008. *Targeted Killing in International Law*, Oxford: Oxford University Press.
Müller, H, and Schörnig, N. 2010. Drohnenkrieg. Die konsequente Fortsetzung der westlichen Revolution in Military Affairs. *Aus Politik und Zeitgeschichte* 50: 16–23.

Müller, R. 2014. Mit Verantwortung. *Frankfurter Allgemeine Zeitung*, June 30. Available online at http://www.faz.net/aktuell/politik/debatte-ueber-kampfdrohnen-mit-verantwortung-13018828.html (3 July 2015).

Nagel, T. 1972. War and Massacre. *Philosophy & Public Affairs* 1(2):123–144.

Robinson, P, Garvey, S, and Kessler Ferzan, Ki. 2009. *Criminal Law Conversations*. Oxford: Oxford University Press.

Ropohl, G. 1993. Neue Wege, die Technik zu verantworten, in Hans Lenk and Günter Ropohl (eds), *Technik und Ethik*. Stuttgart: Reclam, 149–176.

Rousseau, JJ. 1997. *The Social Contract and Other Later Political Writings*. Edited and translated by Victor Gourevitch. Cambridge: Cambridge University Press.

Ryan, C. 2013. Pacifism, Just War, and Self-Defense. *Philosophia* 41: 977–1005.

Shaw, M. 2005. *The Western Way of War*. Cambridge: Polity.

Springer, PJ. 2013. *Military Robots and Drones. A Reference Handbook*. Santa Barbara, CA: ABC-Clio.

Steinhoff, U. 2013. Killing Them Safely. Extreme Asymmetry and Its Discontents, in Bradley Jay Strawser (ed), *Killing by Remote Control. The Ethics of an Unmanned Military*. Oxford 179–208.

The Stimson Center. 2014. *Recommendations and Report of the Task Force on US Drone Policy*. Washington, DC, June. Available online at http://www.stimson.org/images/uploads/task_force_report_final_web_062414.pdf (3 July 2015).

Strawser, BJ. 2010. Moral Predators. The Duty to Employ Uninhabited Aerial Vehicles. *Journal of Military Ethics* 9(4): 342–368.

Strawser, BJ. 2013. *Killing By Remote Control. The Ethics of an Unmanned Military*. Oxford: Oxford University Press.

Talmon, S. 2014. Die Pflicht zur Drohne. Die Menschenrechtskonvention verlangt: Deutschland muss seine Soldaten durch angemessene Ausrüstung schützen. *Frankfurter Allgemeine Zeitung* 10 July 2014, Nr. 157, 7. Available online at http://www.faz.net/aktuell/politik/staat-und-recht/menschenrechte-die-pflicht-zur-drohne-13036828.html (3 July 2015).

Waldron, J. 2010. *Torture, Terror, and Trade-Offs. Philosophy for the White House*. Oxford: Oxford University Press.

Walzer, M. 2004. Two Kinds of Military Responsibility, in Michael Walzer (ed), *Arguing About War*. New Haven/London: Yale University Press, 23–32.

Walzer, M. 2006. *Just and Unjust Wars*. New York: Basic Books.

Werner, MH. 2013. Verantwortung, in Armin Grunwald (ed), *Handbuch Technikethik*. Stuttgart Weimar: J.B. Metzler, 38–43.

Williams, B. 1973. A Critique of Utilitarianism, in JJC Smart and Bernard Williams (eds), *Utilitarianism for and Against*. Cambridge: Cambridge University Press, 77–150.

5 State Responsibility and Drone Operators[1]

Jesse Kirkpatrick

1. Introduction

This chapter explores the responsibilities states have to their members of the military who operate remotely piloted aircraft (RPA) in combat.[2] The ethics of RPA operation – known colloquially as drones – has been a hot area of debate in the past few years. Just war theorists, political scientists, and scholars of public policy have explored topics such as drones and asymmetry (Kahn 2002; Sparrow 2011; Steinhoff 2013), the moral duty to employ drones (Strawser 2010), and the strategic implications of drone use (Boyle 2013).

The most attention has been paid to the U.S.'s drone policy as it relates to unintended killings of noncombatants (Holewinski and Lewis 2013). One reason for this focus stems from the belief that even when such unintended deaths result from lawful combat activity, these deaths can still give rise to various duties for the parties responsible for the killing. Those who subscribe to this belief argue that in the case of civilian casualties, a state has, at the very least, two correlative responsibilities. The first responsibility is to simply acknowledge one's role in bringing about the harm caused, and the second responsibility is to provide a reasonably transparent account of how and why such harm arose.[3] Despite the fruitful scholarship that the debate on states' responsibility vis-à-vis drone operation and civilian casualties has produced, there is another class of agents to whom the state owes significant duties and who have been overlooked largely by scholars, the popular press, and the public: the operators of drones. This chapter aims to fill this gap by exploring the duties that a state and its citizens owe to drone operators.[4]

This chapter proceeds in five parts. This section introduces the chapter. Section 2 discusses why a democratic state has responsibilities to its combat personnel. Section 3 explores drone operation as a form of combat. This section argues that, despite some novel features, drone operation shares important similarities with traditional forms of modern warfare. Section 4 draws from the account of state responsibility developed in Section 2 in order to explicate the duties a state has to combat drone operators. Section 5 concludes.

2. The View of the State[5]

In the past few decades, the dominant normative view of the state found in liberal political philosophy is that it should exist for the mutual advantage of its members

(Gauthier 1986). This view has deep historical roots, which date back to a much earlier period in the history of liberalism.[6] This understanding of the state, often termed in the contemporary literature as the "mutual advantage view," was importantly innovative when it first emerged.[7] No longer was the individual considered inferiorly subordinate to the political leadership; the days when it was taken for granted that people were mere vassals to or subjects of their rulers were eclipsed by the belief that the state existed for the people, not that the people existed for the state.

Within contemporary political philosophy, the mutual advantage view is affiliated with a contractarian understanding of the state. It comes as no surprise, then, that many contemporary social contract theorists also possess the correlative belief that, insofar as individuals are members of the state and to the extent that the state exists for the mutual advantage of these citizens, then there exist some forms of mutual obligation that citizens hold with respect to one another. Subscribing to the mutual advantage view does not commit one to a determinative position regarding how we are to define the various mutual obligations that citizens may hold to one another; nor does it require a specific view concerning how the government may carry out these obligations. Put another way, even if we endorse the mutual advantage view, it remains to be seen just what constitutes "advantage," nor is it clear how this advantage should be effectuated and discharged.

Despite the accommodation for a relatively pluralistic account of how one defines the various forms of mutual obligation that citizens hold with respect to one another, as well as how the government fulfills its attendant fiduciary duties, it is a relatively uncontroversial proposition that the state has substantial responsibilities to categories of individuals who occupy certain roles within society. It is also a commonplace that entailed in these obligations is the commonsense view that the content of particular responsibilities is contingent upon the fact of to whom they are owed. In short, the state has different duties to different people depending on their professions, roles, and so forth. Some of the clearest and most important responsibilities correspond to the people who voluntarily put themselves in harm's way for the protection and defense of their compatriots. Such individuals include police officers, firefighters, emergency personnel, and, perhaps most importantly, members of the military.

There is nothing particularly novel about the argument that the state has various responsibilities to members of the military and that it is the responsibility of government to ensure that these duties are fulfilled.[8] This familiar argument is often brought into sharpest relief when these duties are not met, especially during the times when those in positions of leadership and responsibility fail to provide the adequate equipment, training, oversight, and care to those under their supervision or command.[9] One need only reflect on the recent Veterans Health Administration (VA) scandal in the United States, where significant delays in treatment and care of soldiers within the VA system were considered (correctly) to be an unacceptable breach of the duty to provide medical care to soldiers. Sensing the gravity of this breach, VA employees manipulated data to give the appearance that veterans had shorter wait times for medical care than they actually did. Consequently, numerous veterans suffered from delays in mental and physical care, and it has been alleged that many veterans died during the prolonged wait for their medical care (Oppel 2014).

An even clearer example of the relationship between state and soldier exists in the case of the New Veteran Charter in Canada. The 2005, Charter was an attempt by the federal government of Canada to revise the compensation scheme provided to wounded Canadian soldiers. In turn, a class action lawsuit was filed against Canada's federal government by a group of Canadian veterans. The veterans claimed in their filing that the government had broken the social contract between soldiers and the Crown (i.e. the state), thus abrogating the fiduciary duties associated with the contract. The government countered that no such contract exists between the Crown and the soldiers and that, therefore, the government has no binding fiduciary duties to the soldiers.[10] While the details of the case are complicated, the New Veteran Charter is important because of the way in which the social contract view of the state was invoked explicitly. The Canadian veterans' claim hinged on the notion that such a contract exists and that, consequently, it gives rise to attendant responsibilities that the state owes vis-à-vis its soldiers.

Although this chapter proceeds with a contractarian mutual advantage view of the state and government, addressing issues like those illuminated by the foregoing examples are outside the scope of this chapter. Instead, I wish to endorse a more limited and generally uncontroversial proposition: the state has an obligation to ensure that its duties owed to members of the military are fulfilled.[11] Let us call this *duties to combatants*.[12]

3. Combat Drone Operation and Duties to Combatants: Old Wine in New Bottles?

If we are to take the limited mutual advantage view that I have sketched thus far, how might we think about the relationship between the duties to combatants and combat drone operation? In answering this question, we might first begin by exploring the types of duties that are owed to combatants in general. These general duties may include obligations to provide adequate training, shelter, nutrition, and so forth. While this approach helps establish a general baseline of duties that will be owed to all combatants, it says little about the appreciable differences between, say, an Air Force logistics officer and an Air Force drone operator. Nor does it help us explore presumptive variations in the correlative duties that arise from these differences. Just as there will be some apparent similarities between what is owed to a beat cop and a member of a SWAT team, there will also be some obvious differences that arise from the dissimilarities in these two professional roles. So while this approach may provide us with a very rough outline of duties owed to drone operators *qua* combatants, it fails to provide an answer to the more detailed question concerning responsibility to drone operators *qua* drone operators.

Alternatively, we may choose to explore the specific duties owed to drone operators. One way to proceed with this inquiry is not through abandoning a comparative perspective altogether but by sharpening our units of comparison. In implementing this approach, we can explore how combat drone operation may be similar to other combat roles found in the military. Since we are interested specifically in the kinds of duties that are owed to those involved in *combat* drone operation, in my view, this second analogic approach is the most sensible and efficacious way to proceed; it allows us to tease out the ways that drone operation and their operators may be different from and similar to other forms of combat and combatants and then theorize the extant duties owed to drone operators accordingly.

One challenge to this approach is that it may rest on a false analogy. If drone operation is appreciably dissimilar enough from traditional combat, then the "old-fashioned" *duties to combatants* may be different for drone operators. In fact, many scholars and commentators *do* argue that drone operation is distinctive from the forms of combat that preceded it and that, consequently, drone operators are in turn different from combatants of the past. Those subscribing to this view typically point to three ways in which drones are, at least ostensibly, different from the kind of warfighting that preceded their advent.

3.1 Drones, Casualties, and Democratic Oversight

One potential difference that has appropriately received significant attention is the role that drones play in civilian casualties. The expectation is that compared to most airpower used previously, drones represent a significant increase in technological capabilities that could (and should) lead to better accuracy in targeting. The logical presumption is that this increased accuracy will, in turn, lead to a reduction in noncombatant casualties. The problem is that the United States – the country with one of the largest and most aggressive drone programs – shields civilian casualty counts from the public when they occur in geographic areas like Pakistan and Yemen.[13] The result of this opacity is that it is difficult to assess the competing claims surrounding civilian casualties. Some argue that drone technology produces more accurate and discriminate targeting, thus resulting in reduced civilian casualties, while others argue that drones may be no more discriminate than "manned" airpower and in some cases may even actually *increase* civilian casualties.[14] Absent a substantive change in policy that tends towards greater transparency, we are faced with the prospect of assessing these conflicting arguments based upon incomplete information.

While a military engaging in covert operations is not a novel development, drones may be different in that they increase the ability to kill covertly, and drone operators, of course, have a hand in performing these secretive targeted kills. This lack of transparency is problematic for two interrelated reasons. First, as already mentioned, it presents a significant barrier to account publicly for who is harmed by drone strikes. The issue here is that except for very rare exceptions, there is a duty to acknowledge responsibility for causing collateral damage and, in some cases, even to offer amends for such unintended killings.[15] A second obvious but striking point is that the policies of opacity do not permit the people of a state proper oversight of such killing, a fact that is a considerable source of concern for those who believe that such secrecy is anathema to a properly functioning democratic regime. Without some measure of access to this information, the *people* are left reliant on the goodwill of the government. And given the records of governments, this is a likely recipe for servility.[16]

3.2 Drones and Asymmetry

In addition to the lack of transparency that stems from an unnecessary and overly secretive set of policies, a further set of differences originates from the new warfighting capabilities (and liabilities) that drone technology has made possible. A point of hot debate surrounds the fact that drones increase the asymmetry between those parties who possess them and those who do not. Some view such asymmetric killing as deeply unsettling, reducing war to shooting fish in a barrel and therefore patently unfair (Enemark 2014; Galliott 2012; Kahn 2002). Others who reject such a "boxing match" model of war as a fair fight between opponents on an equal footing view asymmetry as a potential good, especially (and sometimes only) when it benefits a cause that is just (Strawser 2010).

The desire to create asymmetric conditions, particularly asymmetry in the ability to kill, is not a novel development in the history of warfare. Ever since humans have waged battle, there has been a desire to exploit enemy weakness, often through leveraging superior technology that puts combatants at a distance from their adversary. This is true of the catapult, Scud missile, and other "over the horizon" kinds of military weaponry, some of which can now place combatants within striking distance even when they are thousands of miles away from their adversaries.

Aware of this historical fact, both critics and proponents alike are quick to note that, unlike other forms of long-distance weaponry, drones are a game changer because they lower the risks to those who pilot them to nearly zero – a capability, they argue, that is largely unprecedented.[17] Even if contemporary technologies that preceded drones do not require the direct physical location of combatants on the front line, they still necessitate some kinds of physical proximity that can put soldiers at relative degrees of physical risk. The novelty of drones is that combatants can now be completely physically removed from the battlespace while maintaining a high degree of intimacy between the drone operators and their targets (Kirkpatrick 2015). Drone operators now operate systems that allow them to kill by remote control, often from the safety of bases located thousands of miles from the physical war zone.

3.3 Drones as "Riskless" Combat

The final difference between drone operation and previous forms of warfighting is related to the asymmetry objection, and it centers on the idea of drone operation as a riskless endeavor for those who do the operating. Consequently, the argument that drone operation is riskless is held widely.[18] Charles Kels, a major in the Air

Force Reserve and an attorney for the Department of Homeland Security, refers to drone warfare as "a model of warfare that eschews any risk of harm to one party," which leads Kels to conclude that drone operation is a kind of "riskless warfare" (Kels 2012, 223). Political scientist P. W. Singer contends, "[i]f you are fighting from a computer far from the front line, there is no real threat other than carpal tunnel syndrome" (Singer 2013, 15 February). Leading just war theorist Michael Walzer argues similarly that drone technology "makes our enemies more vulnerable than ever before, and we can get at them without any risk to our own soldiers" (Walzer 2013, 11 January). These criticisms are directed at what many call the "Playstation" generation of combatants, who, critics suggest, are more deserving of a decoration for videogaming than one earned in combat.[19]

In addition to the worries of asymmetry associated with "riskless" warfare, an additional criticism of drone technology rests on the notion that drone operators are not engaged in combat and that it would unrealistically stretch the term "combat" if it were to include drone operation. This belief often manifests in the highest ranks of the U.S. military establishment. For example, in his memorandum cancelling the so-called Nintendo Medal – the combat medal for drone operators previously authorized by former Defense Secretary Panetta – former U.S. Defense Secretary Hagel wrote that combat medals should be reserved for those "who incur the . . . hardship of combat," thereby excluding drone operators on the logic that operating a drone does not constitute participation in combat (Hagel 2013, April 15).

The belief that drone operation is sufficiently different from previous forms of combat may cause some to worry about the effects of this "riskless" form of killing. One important concern is predicated on the presumption that the corollary effect of such riskless drone operation on drone operators is the erosion, reduction, and even absence of the important qualities and characteristics that make someone a good and effective combatant. The "stuff" of combatantcy – risk, courage, camaraderie, sacrifice, and so forth – are not and cannot be present.

Ignoring for the moment that numerous drones are actually operated in theater, what are we to make of these claims and criticisms of "riskless" warfare? I have argued elsewhere that such arguments are predicated on truncated and incorrect conceptualizations of risk and injury (Kirkpatrick 2015). Simply put, those who argue that drone operation is riskless ignore or are unaware of the empirical data that proves otherwise. One type of risk that these critics overlook is the risk of psychological injury, namely, post-traumatic stress (PTS).[20] Reliable data on the relationship between RPA operation and psychological injury are growing. For example, a groundbreaking study published in 2013 compared the mental health of RPA operators to that of those operating "manned" aircraft. The study's investigators found "no significant difference in the rates of MH diagnoses, including post-traumatic stress disorder, depressive disorders, and anxiety disorders between RPA and MA pilots" who served in Iraq and Afghanistan at the same time (Otto and Webber 2013, 3).

If it is true that RPA operators do face the risk of psychological injury and suffer from PTS at similar rates as their fellow combatants, why do some scholars contend that drone operation is risk-free? While it is true that such critics may be unaware of the relationship between PTS and drone operation, many of these characterizations are predicated on knee-jerk biases and narrow definitions of risk that focus only on risks to the body. This is a significant disservice to those who face psychological risk and who may also suffer from psychological injury, injury so real that it can be as crippling as a physical injury.

Drone operators face a second kind of risk that may help us understand how drone operation may be different from other forms of combat: the risk of moral injury. Moral injury is commonly defined as "changes in biological, psychological, social, or spiritual functioning resulting from witnessing or perpetrating acts or failures to act that transgress deeply held, communally shared moral beliefs and expectations" (Nash et al. 2010: 1676). With the ability to loiter over targets for days on end, it should come as no surprise that drone operators can also face moral injury. With such up-close and personal surveillance, drone operators often get to "know" their targets. They watch people attend church, buy cigarettes, make love, and kiss their children. They also watch them plant bombs and execute women, and, after lengthy surveillance and familiarity with their targets' patterns of life, drone operators are then called upon to kill them.

Jean Lin Otto, the co-author of the PTS study, notes that drone operators witness "traumatic experiences, such as those associated with PTSD in traditional combat" (Otto and Webber 2013: 3). Such events can include experiencing the aftermath of battle, which "may produce considerable lasting distress and inner turmoil, comparable to consequences of direct life threat" (Litz et al. 2009: 696). Mark Coeckelbergh rightly points out how the power of drones' surveillance technology can shrink the distance between drone operators and those they kill in a way that has deleterious effects on the operators (Coeckelbergh 2013). Although the relationship between drone operation and moral injury has yet to be empirically explored, the prevalence of PTS in drone operators is a good indication that operators could also face the risk of moral injury.

So what are we to make of the differences and similarities we find when comparing drone operation with more traditional forms of combat? And if drone operation does have important differences from other forms of combat, as critics and supporters contend, why have not the duties to drone operators been explored? It seems the putative differences between drone operation and more traditional forms of combat may be one reason the relationship between state responsibility and drone operators has been unexplored. To wit, given the host of issues that ensue from current drone usage and policy, it is perhaps unsurprising that the focus in the literature has been on wrestling with the aforementioned empirical, legal, and normative issues. These are pressing issues, indeed, and scholars have tackled what they believe to be, in terms of importance, those of highest rank.

Some may note in response to this claim that the duties related to drone operators *have* been discussed and that there is an area in the literature positing the possibility that there might be an obligation to use drones (Strawser 2010). But recall, these arguments are advanced in favor of general force protection – they are not obligations to drone operators per se but to soldiers who would be placed in harm's way if drones were *not* used.[21] Such proponents appropriately view drones as tools that increase force protection, but one could very well imagine a similar argument used to advance the duty to use armored vehicles or to provide body armor to troops. While these scholars do discuss state duties, they focus on responsibility vis-à-vis drone *operation*, not vis-à-vis drone *operators*. So not unlike those who seek to press states (i.e. the United States) to fulfill their responsibilities to the parties they harm, advocates of using drones say little about the drone operators themselves. This focus does not, of course, suggest that these arguments reject specific duties to drone operators. I am merely suggesting that these duties have heretofore been unexamined.

4. Duties to Drone Operators

As we have seen, although drone operation is different in some respects from previous forms of combat, it also shares similarities with elements of "traditional" combat. These features are sufficiently similar that we could draw some rough conclusions concerning the state's duties to drone operators by analogically examining the *duties to combatants*. But instead of offering an exhaustive, enumerated list of the *duties to combatants* and comparing and contrasting them to *duties to drone operators*, I will focus on three specific duties to drone operators that are, in my view, the most pertinent.

4.1 Duties to Drone Operators: Recognition

To begin with, drone operation should be recognized for what it is: a form of combat that reduces some risks to drone pilots but that is *not* altogether riskless.[22] Without a clear understanding of how drone operation functions, as well as the risks to operators entailed in the operation of drones, it is hard to see how members of a state can come to sound conclusions concerning what drone operators are owed.

Although recognition may seem relatively simple and straightforward, it is not. While it is true that drones and their operators have been the object of significant public debate, much of it has been entwined with U.S. drone policy. It is no secret that these policies have caused significant consternation internationally, while, conversely, they have been (conditionally) endorsed widely in the United States.[23] And even though polling on public opinion of drone policy has been consistently administered, there is a paucity of data on the public's opinion of drone operators. However, anecdotal evidence suggests that even though members of the U.S. public support drone operations, they still have a low opinion of those doing the operating.

It seems likely that the low standing of drone operators in the public's eye is, at least in part, a consequence of the lack of recognition (i.e. awareness) of what is entailed in drone operation.[24] But even *if* members of the state should recognize the empirical evidence supporting the proposition that drone operation is a form of combat with associated risks and sacrifices, why is this recognition important?

First, such recognition would go a long way in reducing the stigma attached to drone operators. From the charge that drone operators are simply cowards who shrink from a fair fight to the characterization of drone operators as video gaming geeks who can just as easily kill a real human being as they can "kill" a bad guy in a video game, drone operators simply get no respect. Removing this stigma may be instrumentally important in helping to reduce the susceptibility and effects of the psychological and moral injuries that drone operators face and are often afflicted with. As the psychologist Michael Matthews points out, recognition and understanding can "provide a critical protective factor" because "having a network of peers, friends, and family that understand and appreciate the value of the pilot's work helps minimize guilt and moral injury" (Matthews 2014, 55). Matthews' linking people's understanding of the demands of the job and the minimization of psychological and moral harm to drone pilots comports with the folk sense of work, value, and understanding. Anyone who has ever done a difficult or unpleasant job knows that when they describe the demands and hardships of their work, in order for one to feel as if others value and grasp the demands and disagreeable nature of the work, they must first have a rough understanding of what the work entails. The work associated with drone operation is likely no different.

4.2 Duties to Drone Operators: Support?

Matthews also notes how the importance of this recognition can extend beyond pilots' immediate interpersonal relations to include "the support of their fellow citizens" (Matthews 2014, 55). Such support can play a crucial role in helping drone operators know that their sacrifices and contributions to a war effort are valued and recognized. But even if we agree with Matthews' call for the public's greater understanding of the general risks and demands of drone operation, what are we to think about his calls to "support" and "appreciate the value of the pilot's work" (Matthews 2014, 55)?

The answer to this question will depend on how we define "support." Support may include such relatively uncontroversial activities, as Matthews suggests elsewhere in his article, as proper training, screening to identify those most susceptible to PTS, and adequate medical care for the kinds of injuries (moral and psychological) that pose the greatest risk to drone operators (Matthews 2014, 54–55).[25] Let us call support of activities like these "narrow support." *Narrow support* is limited to supporting general physical and psychological needs by way of goods and services that are typical for combatants and those goods and services (like those aforementioned) specific to drone operators.

While we may agree that a state has a duty to offer narrow support to drone operators, such support should not be equated with the more expansive notion of support for specific drone operations and the operators who carry them out. This rejection of a default to the more robust view of support for specific operations and operators – let us call it *broad support* – rests upon key elements of the revisionist view of just war theory. The orthodox view of just war theory holds that two sets of principles constitute the morality of war: those informing the resort to war (*jus ad bellum*) and those informing the conduct in war (*jus in bello*). A commonplace of traditional just war theory is the view that the *in bello* rules of war are "logically independent" from the *ad bellum* rules (Walzer 2006, 20). Accordingly, orthodox just war theorists argue that while soldiers must follow the rules and norms informing *in bello* conduct, they are not barred from fighting on behalf of an unjust cause. Conversely, the revisionist camp of just war theory denies the logical separation of *ad bellum* and *in bello*. Whereas orthodox just war theory sees combatants as morally equal irrespective of the justice of their cause, revisionists reject this equality, arguing that combatants with an objectively unjust cause (unjust combatants) have no moral right to fight.

The point here is not to intimate that those who subscribe to the orthodox view of just war theory are not and cannot be critical of drone operators' *in bello* conduct or of the *ad bellum* cause they serve. And while some may value and support their state's combatants in the abstract, they may also disvalue and oppose the activities of those soldiers who fight for an unjust cause. But the orthodox approach does permit the view that drone operators are simply a subclass of combatants who, not unlike combatants *writ large*, deserve the *broad support* of members of the state irrespective of the justice of their cause. Unlike orthodox just war theory, under the revisionist account, such *broad support* of drone operators is necessarily contingent upon the justice of the cause for which drone operators serve. So while neither orthodox nor revisionist just war theory claims that a just cause is sufficient for *broad support*, just cause is necessary for revisionists to engage in *broad support*, a condition that is absent in orthodox just war theory.

Under the revisionist view, we should, then, reject the idea that *broad support* can be a duty, for how could it be a duty of the state to support war efforts in a general, whole cloth fashion? And even making broad support a conditional duty contingent on the justice of the cause goes too far because there are numerous members of the state who may reasonably reject war altogether, and they should not be thought of as in breach of their duty when they do so.

4.3 Duties to Drone Operators: Transparency

No matter which strand of just war theory one subscribes to – revisionist or orthodox – the notion of *broad support* points to the crucial importance of reasonable transparency in drone operations. Transparency allows members of the state to make informed decisions concerning *broad support*: namely, whether they do or do not support the war effort, the actions used to achieve the effort (i.e. drone operation), and those engaged in the actions (i.e. drone operators). Transparency need not involve the minutiae of operational details but must include the broader context of identifying clearly the enemy,[26] providing publically available legal and moral rationales for targeting the enemy,[27] criteria for target selection, and the accounting of and remedies for civilian casualties.[28]

In addition to providing information that is so important to citizens' deliberative processes involved in determining their positions on *broad support,* this transparency can also be thought of as a duty to drone operators. I am referring here to the duty for members of the state to cajole the government to be reasonably open and transparent with respect to drone policy and operations; let us call this the *duty of reasonable transparency*.

The prevailing view in most democratic countries is that the military is subordinate to civilian control.[29] This notion of soldierly obedience, along with the correlative view of the obedient soldier as an upright and decent one, dates back centuries. We see this view in the remarks of U.S. Supreme Court Justice Oliver Wendell Holmes when he remarked, "there is one thing I do not doubt . . . and that is that the faith is true and adorable which leads a soldier to throw away his life in obedience to a blindly accepted duty, in a cause which he little understands" (Lerner 1943, 20). A similar sentiment can be found in the more contemporary remarks of leading just war theorist, Michael Walzer: "we draw a line between the war itself, for which soldiers are not responsible, and the conduct of the war, for which they are responsible, at least within their own sphere of activity" (Walzer 2006, 38).

A cornerstone of this model of civil–military relations is that in democracies, the citizenry are responsible for ensuring the military actions carried out in their name and for their defense have a just cause. In turn, the military agree to rely on these democratic processes of oversight and participation. Consequently, it is imperative that members of the state play a role in helping to ensure that drone operators are not called upon to serve an unjust cause. Yet this imperative can be fulfilled only if the citizenry have access to reliable and accurate information made possible through reasonable transparency.

In addition to ensuring that the *ad bellum* principle of just cause is satisfied, an active citizenry advocating for reasonably transparent policies can also help ensure that drone policies respect *in bello* just war principles, specifically the principles of distinction, proportionality, and necessity. Here we can see how the call for transparency made so urgently and repeatedly by advocates of noncombatants comes full circle. This is because it is incumbent upon a democratic citizenry to help ensure that its government's policies and practices are the most humane. This is for the sake of both the noncombatants who will be harmed and for those combatants doing the harming. Just as a combatant with any decency wants to avoid harm or injury except when absolutely necessary, with the most judicious and proper discrimination in targeting and in a way that is as proportionate as possible, it is also obligatory for the members of a state to ensure that the proper policies and practices are in place that uphold and respect these *in bello* principles. And when these principles are breached or if noncombatants are harmed in the course of legal, proportionate, necessary, and discriminate *in bello* actions, then a full account and appropriate remedy needs to be provided. Such transparency is owed to both those harmed and those combatants who have done the harming.[30]

5. Conclusion

In this chapter, I have explored the state's *duties to drone operators*. I noted that while others have explored the relationship between duties and drone operation, the primary focus of these explorations has not been on what is owed to the operators. This chapter has sought to fill this gap in the discussion. To wit, I explored the shared similarities and differences between the duties to "traditional" combatants and those owed to drone operators. I, in turn, focused on three duties: (1) the duty to recognize drone operation as a form of combat with its own demands and associated risks, (2) the duty to offer *narrow support* to drone operators, and (3) the duty to create as transparent policies as reasonably practicable to help ensure that drone operators and operations are only used proportionately, when necessary, and in furtherance of a just cause. Given the fact that drone technology increases the ease with which governments can carry out covert operations, including killings of intended and unintended people, this latter duty is particularly important. If we are to believe, as many do, that wars can be just, then it is incumbent upon a state to ensure that its men and women in uniform fight only for a just cause and fight for this cause justly. This is one of the greatest duties a state owes to drone operators.

Notes

1 An early draft of this chapter was written while I was Class of 1978 Resident Research Fellow at the U.S. Naval Academy, Stockdale Center for Ethical Leadership. I am very grateful for the Center's support. A draft of this chapter was presented at the International Studies Association–South 2014 conference and the 2015 Annual Meeting of the Southern Political Science Association. I am thankful for the helpful comments I received.

2 The term "remotely piloted aircraft" (RPA) and the more colloquial "drones" will be used interchangeably throughout this chapter. I view these terms as preferable to "unmanned aerial vehicles" (UAV), which connotes that such systems/aircraft are operated only by men. Furthermore, the term "unmanned" can also give the impression that such aircraft are autonomous or do not require the large number of service members on the ground who both operate these aircraft and are required to support their operation. I use the term "drone operator" to include both the pilot of the drone and the sensor operator (the individual responsible for control of the system's cameras and missile firing and control). I should also add that I am here referring to combat drone operation.

3 With respect to the United States' targeted killing program, the Obama administration's lack of transparency on these fronts has been a source of significant criticism. See Amnesty International et al. (2014).

4 I should add that the state and government used by way of example is the United States. The reason for this should be clear: the United States has one of the largest, most aggressive, and, in the case of targeting killing, most secretive drone programs in the world.

5 The mutual advantage view that I draw from comports with several models of the state. See Buchanan (1999).

6 Perhaps the most persuasive and well-known modern account of this view is found in John Locke's *Second Treatise* (Locke 1991).

7 For a useful discussion on this innovation, see Morris (2002: 6–7).

8 It is important to introduce the distinction between the state and the government, a distinction common among those who hold the mutual advantage view. Where the "state is an association for the mutual advantage of its members . . . the government is simply an agent whose fiduciary duty is to serve the interests, or to realize the will of those citizens" (Buchanan 1999: 73). For a similar view, see Raz (1986: 70).

9 State responsibility should not be confused with reverence or the conferral of exalted status on members of the military. In fact, there are very good reasons, not least among them the grave injustices that members of the military often commit, to be reluctant to resort to such celebrations. But such reluctance is not incompatible with the very limited claim that the state has significant responsibility to its combatants.

10 For a general discussion of the case over the New Veteran Charter, see Everson (2014). Interestingly, the breach of the claimed social covenant was based upon the honor of the Crown, a doctrine that stemmed from the Crown's dealings with aboriginals and held that in such dealings (and perhaps other dealings, e.g. those with Canadian veterans), the Crown must act honorably. See Haida Nation Case (Haida Nation Case 2004, paras. 16–19, 25). It is important to note that the government did not claim that such a contract does not exist in any form, rather that the contract does not exist in a legally binding way (i.e. in written or unwritten statutory, regulatory, or constitutional form).

11 There is room here for debate on how we are to define these responsibilities, an issue I will discuss.

12 I use the term "combatants" generically to include soldiers, marines, airmen and airwomen, and sailors.

13 Thanks to an anonymous reviewer for noting that the veil of secrecy has been lifted for some aspects of the United States' targeted killing program, including the targeting process. Despite some transparency in the program, and although the Obama administration has acknowledged that it uses force in Pakistan and Yemen, the administration continues to say very little officially to the public about its involvement in and facts surrounding targeted killing operations in these countries. For targeting procedures, see United States Department of the Army (2015). For a thorough overview of targeted killing, see McNeal (2014). For U.S. acknowledgment of the use of force in Yemen, see Obama (2012). For a discussion of the continued lack of transparency, see Baker (2015). For an attempt in U.S. Congress to mandate transparency in targeted killing, see Representative Schiff and Adam B. (2014).

14 Figures on civilian casualties resulting from drone strikes are further complicated by the various locations that they occur (e.g. Afghanistan, Pakistan, Yemen, Somalia) and who is performing the strikes (the CIA or U.S. military). For an analysis of figures on CIA strikes in Pakistan and their reduction in civilian casualties, see Plaw, Fricker, and Williams (2011). Holewinski and Lewis claim research supports the conclusion that drone strikes were more likely to cause civilian casualties in Afghanistan than those by "manned" aircraft; they speculatively attribute this to deficiencies in training (Lewis and Holewinski 2013: 60).

15 "Amends" is the emerging idea that civilians harmed in war, even incidentally and legally, are owed assistance. In a rare case of acknowledgment, the Obama administration admitted responsibility for the unintended killing of two hostages held by Al Qaeda in Pakistan. In addition to releasing facts about the case, the administration also committed to paying financial compensation to the victims' families. See http://www.nytimes.com/2015/04/24/world/asia/2-qaeda-hostages-were-accidentally-killed-in-us-raid-white-house-says.html?_r=0.

16 I will return to exploring the relationship between transparency and a state's duty to drone operators in the next section.

17 It should be noted that this asymmetry is not entirely novel. For example, nuclear intercontinental ballistic missiles present a similar type of asymmetry. My assumption is that these when these critics refer to asymmetry, they do so with respect to conventional weapons.

18 Criticisms of new technologies that make long-distance killing possible are not unprecedented; history is replete with examples. As Benjamin Constant, writing in 1814, lamented: "The new way of fighting, the changes in weapons, artillery, have deprived military life of what made it most attractive . . . the development of our physical and moral faculties, that made hand-to-hand fighting so attractive to the

heroes of antiquity or to the knights of the Middle Ages. War has lost its charm" (Constant 1988, 55).

19 See Philip Alston for a discussion of the need to avoid such a "Playstation mentality" in drone operation (Alston 2010, 25).

20 In keeping with some mental health practitioners, I view post-traumatic stress not as a disorder but as a kind of psychological injury. I therefore refer to PTS as opposed to PTSD.

21 Some have also argued there may be a duty to use drones because they increase discrimination and proportionality, thereby reducing civilian casualties. Whatever one thinks of these arguments that rest on empirical claims, they are, strictly speaking, about the duty not to drone operators but instead to those who may be harmed by drone operation. Still, one may argue that there is a secondary duty to combatants in general to provide them with the most humane, discriminate, and accurate weaponry available so as to permit them to soldier in the best (i.e. humane and discriminate) way possible, thus avoiding civilian casualties.

22 I hasten to add that recognition should not be seen as an endorsement of drones, drone operation, or drone policy; it is not. One can at once *merely* recognize drone operation as a form of combat with extant risks, while remaining simultaneously critical or even repulsed by the use of drones. There is a bright distinction between recognition and reverence; I do not mean to imply that there is a duty to support specific or general drone operations and policies but simply to acknowledge the facts of the matter. See note 7 on a similar point. I should also add that recognition is not a duty in the strict sense of the term, as I do not believe that individuals have a duty to acknowledge and accept epistemic facts.

23 I add the modified "conditional" because, as Kreps has demonstrated, support of U.S. drone policy is contingent and significantly varies when more controversial elements of the program are introduced into the survey instruments. Kreps has shown that experimental treatments including the International Humanitarian Law (IHL) principles of proportionality and distinction reduce public support for drone operation and policy (Kreps 2014).

24 This is not to suggest that there are not many who *do* understand the role of drone operators in combat and the risks they face yet still simply believe drone operation is a form of lesser military service than more "traditional" forms of combat.

25 Adequate medical care is particularly important given the stigma that was and in many cases still is attached to PTS as a form of injury.

26 The Obama administration has refused to disclose the "associated forces," that is those forces associated with al-Qaeda that it claims it has the legal authorization to attack. The administration has publicly stated the rationale for this denial of disclosure: that divulging the groups' names could elevate their position and therefore strengthen their standing and credibility. For an excellent discussion on citizenship and judicial review, see Vladeck (2014).

27 These rationales should be made available *ex ante* and not *post facto*, as was the case with the release of the Department of Justice's memo on the legal authority to extrajudicially kill a U.S. citizen. See Isikoff (February 4, 2014).

28 To the extent practicable, citizens should know who has struck and when and why, a matter made difficult by the fact that such strikes are often opposed strongly by the public in the places in which they occur and typically happen with the secret and tacit support of the government.

29 Recall that traditional just war theory holds that because soldiers have little to no control over *why* they fight, such decisions are the domain of the civilian political leadership; they are responsible onlyfor *how* they fight.

30 The duty for reasonable transparency is, of course, owed to all combatants. But given the technological capabilities of drones, their operation can be and has been particularly hidden under a shroud of secrecy; a state owes it to its men and women in uniform for the shroud to be lifted.

References

Alston, P. 2010. *Report of the Special Rapporteur on Extrajudicial, Summary and Arbitrary Executions, Addendum: Study on Targeted Killings*. A/HRC/14/24/Add. 6, 2010. Accessed 27 January 2016. Available online from http://www2.ohchr.org/english/bodies/hrcouncil/docs/14session/A.HRC.14.24.Add6.pdf.

Amnesty International. 2014. *Joint Statement in Support of the Targeted Lethal Force Transparency Act*. Accessed 27 January, 2016. Available online from https://www.hrw.org/news/2014/04/02/joint-statement-support-targeted-lethal-force-transparency-act.

Baker, P. 2015. Obama Apologizes after Drone Kills American and Italian Held by Al Qaeda. *New York Times*. Accessed 27 January, 2016. Available online from http://www.nytimes.com/2015/04/24/world/asia/2-qaeda-hostages-were-accidentally-killed-in-us-raid-white-house-says.html?_r=0.

Boyle, MJ. 2013. The Costs and Consequences of Drone Warfare. *International Affairs* 89(1): 1–29.

Buchanan, A. 1999. The Internal Legitimacy of Humanitarian Intervention. *Journal of Political Philosophy* 7(1): 71–87.

Coeckelbergh, M. 2013. Drones, Information Technology, and Distance: Mapping the Moral Epistemology of Remote Fighting. *Ethics and Information Technology* 15(2): 87–98.

Constant, B. 1988. *"The Spirit of Conquest" in Political Writings*. Cambridge Texts in the History of Political Thought. Cambridge University Press.

Enemark, C. 2014. *Armed Drones and the Ethics of War*. Routledge.

Everson, K. 2014. *Veterans Don't Have Social Contract, Ottawa Says in Lawsuit Response*. Accessed 27 January, 2016. Available online from http://www.cbc.ca/news/politics/veterans-don-t-have-social-contract-ottawa-says-in-lawsuit-response-1.2577053.

Galliott, JC. 2012. Uninhabited Aerial Vehicles and the Asymmetry Objection: A Response to Strawser. *Journal of Military Ethics* 11(1): 58–66.

Gauthier, DP. 1986. *Morals by Agreement*. Oxford University Press.

Hagel, C. 2013. *Memorandum: Distinguished Warfare Medal*. Edited by Office of the Secretary of Defense, April 15. Accessed 27 January, 2016. Available online from http://archive.defense.gov/pubs/PR%20and%20CJCS%20Memo%20DWM.pdf.

Haida Nation Case. 2004. *Haida Nation v. British Columbia (Minister of Forests)*, (2004) 3 S.C.R. 511.

Holewinski, S, and Lewis, L. 2013. *Five Ways Obama Can Fix Drones Right Now*. Accessed 27 January, 2016. Available online from http://www.defenseone.com/ideas/2013/11/five-ways-obama-can-fix-drones-right-now/73304/?oref=d-skybox.

Isikoff, M. 2014. *Justice Department Memo Reveals Legal Case For Drone Strikes on Americans*. 4 February. Accessed 27 January, 2016. Available online at http://investigations.nbcnews.com/_news/2013/02/04/16843014-justice-department-memo-reveals-legal-case-for-drone-strikes-on-americans?lite.

Kahn, PW. 2002. The Paradox of Riskless Warfare. *Philosophy and Public Policy Quarterly* 22(2): 3.

Kels, CG. 2012. Closing Argument: At the Outer Bounds of Asymmetry. *Journal of Military Ethics* 11(3): 223–244.

Kirkpatrick, J. 2015. Drones and the Martial Virtue Courage. *Journal of Military Ethics* 14(3/4): 202–219.

Kreps, S. 2014. Flying under the Radar: A Study of Public Attitudes towards Unmanned Aerial Vehicles. *Research & Politics* 1 (1).

Lerner, M. 1943. *The Mind and Faith of Justice Holmes*. Little, Brown & Company.

Lewis, L, and Holewinski, S. 2013. Changing of the Guard: Civilian Protection for an Evolving Military. *Prism* 4(2): 57–65.

Litz, BT, Stein, N, Delaney, E, Lebowitz, L, Nash, WP, Silva, C, and Maguen, S. 2009. Moral Injury and Moral Repair in War Veterans: A Preliminary Model and Intervention Strategy. *Clinical Psychology Review* 29(8): 695–706.

Locke, John. 1991. *The Second Treatise of Government: An Essay Concerning the True Original, Extent, and End of Civil Government.* Cambridge University Press.

Matthews, MD. 2014. Special: Stress among UAV Operators – Posttraumatic Stress Disorder, Existential Crisis, or Moral Injury? *Ethics and Armed Forces* 1(1): 53–57.

McNeal, GS. 2014. Targeted Killing and Accountability. *The Georgetown Law Journal* 102: 681–794.

Morris, CW. 2002. *An Essay on the Modern State*. Cambridge University Press.

Nash, WP, Vasterling, J, Ewing-Cobbs, L, Horn, S, Gaskin, T, Golden, J, Riley, WT, Bowles, SV, Favret, J, and Lester, P. 2010. Consensus Recommendations for Common Data Elements for Operational Stress Research and Surveillance: Report of a Federal Interagency Working Group. *Archive of Physical Medicine and Rehabilitation* 91(11): 1673–1683.

Obama, B. 2012. *Presidential Letter – 2012 War Powers Resolution 6-Month Report.* Accessed 10 June 2015. Available online at https://www.whitehouse.gov/the-press-office/2012/12/14/letter-president-war-powers-resolution.

Oppel, R. 2014. *Watchdog Says V.A. Officials Lied.* Accessed 27 January 2016. Available online at http://www.nytimes.com/2014/09/10/us/watchdog-says-va-officials-lied.html.

Otto, JL, and Webber, BJ. 2013. Mental Health Diagnoses and Counseling among Pilots of Remotely Piloted Aircraft in the United States Air Force. *MSMR* 20(3): 3–8.

Plaw, A, Fricker, MS, and Williams, BG. 2011. Practice Makes Perfect?: The Changing Civilian Toll of CIA Drone Strikes in Pakistan. *Perspectives on Terrorism* 5(5–6).

Raz, J. 1986. *The Morality of Freedom*. Oxford University Press.

Representative Schiff, and Adam, B. 2014. *H.R.4372-Targeted Lethal Force Transparency Act. 113th U.S. Congress*. Accessed 27 January 2016. Available online at https://www.congress.gov/bill/113th-congress/house-bill/4372.

Singer, PW. 2013. *A Military Medal for Drone Strikes? Makes Sense*. Washington Post Opinions. 15 February 2013. Accessed 10 June 2013. Available online at https://www.washingtonpost.com/opinions/a-military-medal-for-drone-strikes-makes-sense/2013/02/15/e90c0638-76e4-11e2-8f84-3e4b513b1a13_story.html.

Sparrow, R. 2011. Robotic weapons and the future of war, in P Tripodi and J Wolfendale (eds). *New Wars and New Soldiers: Military Ethics in the Contemporary World*, Ashgate, 117–134.

Steinhoff, U. 2013. Killing Them Safely: Extreme Asymmetry and Its Discontents. In BJ Strawser (ed). *Killing by Remote Control*. Oxford University Press, 179–208.

Strawser, Bradley Jay. 2010. Moral Predators: The Duty to Employ Uninhabited Aerial Vehicles. *Journal of Military Ethics* 9(4): 342–368.

United States Department of the Army. 2015. *Army Techniques Publication 3–60. Targeting*. Accessed 27 January 2016. Available online at http://armypubs.army.mil/doctrine/DR_pubs/dr_a/pdf/atp3_60.pdf.

Vladeck, SI. 2014. Targeted Killing and Judicial Review. *George Washington Law Review* 82(11A): 556.

Walzer, M. 2006. *Just and Unjust Wars: A Moral Argument with Historical Illustrations*. 4th ed. Basic Books.

———. 2013. *Targeted Killing and Drone Warfare. January 11, 2013. Dissent*. Accessed 27 January 2016. Available online at http://www.dissentmagazine.org/online_articles/targeted-killing-and-drone-warfare.

Part III

Design and Sociotechnical Perspectives

6 The Threshold of Killing Drones

The Modular Turing Imitation Game

Asa Kasher

1. Introduction

For a while now, a ban has been proposed on the usage of any drone that will be programmed to make autonomous decisions to use its weapons in a direct attempt to kill human beings.[1] The purpose of the present chapter is to show that an absolute ban on such drones would be unreasonable. Our argument is going to rest on a discussion of the possibility of positing a threshold for the usage of such killing drones. On the grounds of Turing's Imitation Game, we propose the use of a Modular Generalized Imitation Game. A suggestion has been made on the grounds of slippery slope arguments to ban usage of drones for killing people even if the decisions to kill are made by human operators. We will argue against such slippery slope arguments as well.

2. The Imitation Game

Our starting point is an interesting aspect of the famous Imitation Game known as the Turing Test.[2]

The Imitation Game involves three players, a human interrogator and two reply sources, one of which is a person and the other one a computer.[3] Steps in the game are simple: the interrogator poses a question to one of the reply sources, without knowing whether the interrogated source is the person or the computer. The reply source replies when interrogated. The point of the game is similar but interestingly different for each of the reply sources: the person intends to try to help the interrogator identify him as the person and not as the computer, whereas the computer is programmed to try to mislead the interrogator into thinking that it is the person and not the computer. The interrogator has to form a view as to which reply source is the person and which is the computer. If the interrogator's view is correct, then the computer has lost and the person won, and if the interrogator's view is incorrect, then the computer has won and the person has lost.

Alan Turing, who suggested the Imitation Game (in Turing 1950), was interested in the possibility that "the use of words and general educated opinion will have altered so much that one will be able to speak of machines thinking without expecting to be contradicted". The leading intuition was that one should not be contradicted when speaking about computers as thinking objects, if computers have been shown to be on a par with human beings, at least with respect to some activities that seem to involve intelligent considerations. In the present chapter, this leading intuition is not going to play a major role. The possibility of a computer acting on a par with human beings is in our focus.

A computer has not been shown to be on a par with human beings in respect of Turing's Imitation Game if it won one round of the game in which it managed to mislead the interrogator into taking it to be the human reply source. The result of a single round reflects the effects of the particular decisions made by each of the three participants during that round. Winning a round of the Imitation Game does not reflect the general ability of the computer to mislead the interrogator. Hence, winning in rounds of the Imitation Game counts as strong evidence for the computer being on a par with human beings in a certain respect only if the computer wins rounds of the Imitation Game regularly rather than sporadically.

Winning regularly does not mean winning each and every round of the Imitation Game. Turing has been interpreted as suggesting a lower threshold for success in rounds of the Imitation Game counting as strong evidence for the computer being on a par with human beings in a certain respect that seems to involve intelligent considerations (Oppy and Dowe 2011). Turing himself said: "I believe that in about fifty years' time it will be possible to programme computers, with a storage capacity of about 10^9, to make them play the Imitation Game so well that an average interrogator will not have more than 70 percent chance of making the right identification after five minutes of questioning". For the sake of our discussion in the sequel, let us define two thresholds. The *minimal threshold* is that of the computer being able to mislead the interrogator in at least 30% of the rounds that are restricted to 5 minutes of interrogation. The *maximal threshold* is that of the computer being able to mislead the interrogator in almost every round of whatever length. The latter threshold is more important for our purposes than the former one.

3. The Modular Generalized Imitation Game

Turing's Imitation Game includes three components. First, there is a computer programmed to successfully carry out, on a regular basis, a task that seems to involve intelligent considerations; second, the performance of the computer is, in some strict sense, on a par with human performance of tasks of the same kind; and third, given the nature of the tasks under consideration and the regular success of the computer, it is reasonable to ascribe to the computer the ability to think.

Much of the philosophical discussion of the Turing Test has been devoted to the third component. We are interested in the other two components. Notice that we have presented the first and the second components in terms that are independent of the third component. The task of the computer is not necessarily to mislead the interrogator by answering questions in a successfully misleading way. The task could serve other purposes, but it is required to seemingly involve intelligent considerations and to be regularly successful. Moreover, the task does not have to be confined to answering the questions of the interrogator. It could involve any performance that the interrogator can be aware of and understand. The task of the human participant too does not have to be answering the questions of the interrogator. It could be any task that, first, the interrogator can be aware of and understand, and second, is sufficiently similar to the task of the computer to justify a comparison of their rates of success in performing their tasks. Most interesting would be the case of the human participant and the computer having to perform exactly the same task.

Here, then, is the Modular Generalized Imitation Game we would like to consider.[4]

The participants are a human being, a computer, and a human evaluator who is an expert in a certain area of knowledge and activity. Let us dub this area of expertise "the problem area".

In each round of the game, the evaluator presents the participants with a problem from the problem area that they have to solve within a specified period of time. The task seems to involve intelligent considerations. When the time elapses, the participants present what they describe as their solutions. On the grounds of personal expertise and the given solutions, the evaluator has to make four decisions: first, whether the solution suggested by the human participant is an adequate solution; second, whether the solution suggested by the computer is an adequate solution; third, if the suggested solutions are adequate, whether one of them is better than the other; and fourth, which suggested solution was provided by the human participant and which by the computer.

The point of the game for the human participant and the computer is to solve the given problem most adequately. The point of the game for the evaluator is to do his best in evaluating the suggested solutions and in trying to identify the two other participants on the grounds of their suggested solutions.

A point is scored by the human participant and the computer every time one's suggested solution is found by the evaluator to be adequate. Another point is scored by one of them if one's suggested solution is considered by the evaluator to be better than the other suggested solution. The winner among the two is the one who scored more than the other.

4. Minimal and Maximal Thresholds

The role played by the evaluator in the Modular Generalized Imitation Game, when she tries to identify the computer among the two participants, is actually the role of the interrogator in the original Imitation Game. The more each of them is unsuccessful in identifying the computer among the participants, the more successful is the performance of the computer in solving the given problems, which seems to involve intelligent considerations.

Turing suggested that a certain rate of success of the computer in the Imitation Game would count as firm grounds for ascribing to the computer the capacity of thinking, in the sense of the term used in describing human beings as thinking. Similarly, one can put the Modular Generalized Imitation Game to such a use.

If the problem the computer has to solve, within the framework of the Modular Generalized Imitation Game, is, say, to identify objects of a certain nature in an aerial photograph, produce an abstract of a one-page-long text, or decipher an acronym, then a certain rate of repeated failure on the part of the evaluator in identifying the computer among the participants would justify ascribing the computer the capacity of identifying, abstracting, or deciphering, all of which seem to involve intelligent considerations.

Assume we have been convinced by the performance of the computer that it should be ascribed the related capacity. Would it be reasonable to use it on a regular basis to provide us with its solutions to given problems in its area? Would it be reasonable to use it as our only source of such solutions, replacing all human beings who used to be our reliable sources of solutions?

At this point, our distinction between a minimal threshold and a maximal one seems useful. The minimal threshold suggested by Turing's paper should be used only for ascribing a certain capacity to the computer under consideration. The moves from ascription to regular usage and then to regular exclusive usage will be reasonable only if the quality of the computer is not only above the minimal threshold but also above some higher threshold. How much higher should such a higher threshold be depends on the nature of the task performed by the computer, that is to say, the nature of the use to which the solutions provided by the computer are going to be regularly put. If the abstract produced by a programmed abstractor is going to serve as a mere suggestion proposed to readers of some Internet pages on entertainment, then perhaps the higher threshold does not have to be much above the minimal threshold. However, if the deciphered acronyms are going to serve some legal or another formal purpose, then a much higher threshold would naturally be required.

How much higher should a higher threshold be? Should it be the maximal threshold? Recall that Turing's Imitation Game and our Modular Generalized Imitation Game involve a comparison of the computer performance in solving problems of a certain kind with that of a human being assigned the same problems. The quality of human performance serves as the standard. The closer the computer's performance is to this standard, the more successful it is taken to be. Hence, setting a higher threshold for regular usage of a computer as a source of solutions of given problems depends on the extent to which the computer's performance is similar to the human one.

Notice that the quality of human performance is our standard, though it may well leave room for improvement.[5] We regularly use products of human performance that are adequate but far from being of very high quality because there is nothing at stake that would require a much higher quality. Compare, for example, the standards of accuracy required when medical prescriptions are under consideration with the standards required when an entertainment page of a daily newspaper is being read. Generally, we assume, for the purpose of discussing the Imitation Game, the Modular Generalized Imitation Game, and similar ones, that the standards of human performance are often adequate and acceptable, even if improvable.

5. The Drone Game

We return now to the issues of killing drones. Our Drone Game is a Modular Generalized Imitation Game of the kind we discussed in the previous sections, where the task assigned to the human participant and the computer is that of guiding a drone of a certain kind under given circumstances that very much resemble the circumstances of real combat operations in which drones play a major role.

The duty of the evaluator within the framework of the Drone Game is, first, to evaluate the performance of the other participants and see whether their actions are adequate, and second, if their actions are adequate, determine which of them should be considered as professionally better. As usual, the evaluator will also have to try to identify the computer among the participants.

Now, let us consider a possible evaluation of the participants' performance after a long series of rounds of the Drone Game that involved simulated circumstances of all the kinds of real circumstances in which drones are involved in combat. Assume that the evaluation is (a) that both participants' actions are almost always adequate from a strict professional point of view and (b) that the participants are indistinguishable from each other, which means that the performance of the computer is on a par with that of a human operator. Notice that such an evaluation meets our definition of the maximal threshold.

Here is the major moral question with respect to killing drones, ones that, on the grounds of the given data about the given target and what is in its vicinity, make autonomous decisions to kill and employ their weapons in order to carry out those decisions. On the background of the given evaluation, will the usage of such drones be morally justifiable?

The burden of proof rests with those who would like to claim that the usage of such drones is morally unjustifiable. If a computer acts under some circumstances exactly as human operators act under the same circumstances and the latter's acts are morally evaluated as adequate, that is to say justifiable, then why shouldn't the usage of the former's acts also be morally justifiable? We will, therefore, consider some arguments that try to show that the usage of such killing drones is morally unjustifiable and see whether they hold water.

The first argument to be considered is the Error Argument. Its premises are reasonable: in decision-making with respect to matters of life and death, errors should be avoided to the greatest extent possible under the circumstances, and errors are bound to occur come what may. In order to draw the conclusion, when matters of life and death are at stake, that human operators should be in charge rather than computers, additional premises have to be used, namely (a) that human beings are less prone than computers to commit significant errors and (b) that human beings are better than every computer in performing tasks of error detection and error correction.

Recall that the computer under discussion is one that has emerged as indistinguishable from a human operator in an adequately long series of rounds of the Drone Game. Accordingly, one may assume that both the human operator and the computer are indistinguishable from each other in respect of their proclivity for making decisions that we consider to be erroneous. Hence, the Error Argument involves only extraordinarily rare errors, ones that were not detected during the long series rounds of the Drone Game test.

Do we have reason to assume that human operators will commit fewer errors than a computer when they have to quickly solve an extremely surprising problem of an utterly new type? Assume the evaluation of the performance of both the human operator and the computer included an evaluation of how each of them solves new problems when they first encounter them. The final evaluation of them as indistinguishable from each other means that they are on a par with each other in that respect too. If there is a reason for some marginal preference of one of them over the other, it would be preference of the computer because it is never tired, angry, or panicked.

We assume that the advanced systems of human operator and computer include components of self-control that try to detect errors just committed and appropriately correct them. Again, since the systems are indistinguishable from each other, there is no reason to assume that one of them is better than the other in respect of error detection or error correction.

The next argument we consider is the Human Dignity Argument. It rests on some conception of what it means to respect human dignity and draws the conclusion that computers as such, even if programmed to be indistinguishable from human operators in performing certain tasks, should not be taken to be able to properly respect human dignity. Their moral shortcoming is reflected not in their performance but in their essence as computers rather than persons.

The Human Dignity Argument is actually directed against whoever has made the decision to use a computer rather than a human operator for targeted killing or similar military purposes. Such a decision-maker has not shown proper respect for human dignity by allowing usage of a computer for those purposes. I beg to differ. In order to manifest in one's decisions proper respect for human dignity, a decision-maker ought to act in a way that (a) reflects a positive attitude towards human beings as such; (b) does it in a way that conspicuously marks a difference between human beings and tools and other artefacts as well as familiar non-human beings; and, most importantly, (c) is governed by principles of respect for the personhood of human beings – that is their reason, will, and freedom – by imposing restrictions on every action and method to be used that has or may have an impact on the extent of liberty that persons enjoy. A decision to use a computer under the present circumstances seems to us to be compatible with all three requirements.

We assume that a human operator's performance has been evaluated as adequate only if it regularly reflects a positive attitude towards human beings as such. An operator reflects in one's activity such an attitude on the grounds of general conceptions, principles, and procedures that shape and guide the operator's activity. Given that the computer is indistinguishable from the human operator in its activity within the problem area, it is clear that the computer is programmed in a way that embodies equivalent general conceptions, principles, and procedures, perhaps even the same ones, and therefore acts in a way that manifests the same positive attitude. Strictly speaking, a computer does not have an attitude, but what we mean is indeed that if we interpret the computer's activity as if it is human activity, then the natural interpretation of its activity would be that the computer shows a positive attitude towards human beings as such. Thus, the computer under consideration satisfies the necessary condition (a). A similar consideration shows that it satisfies necessary condition (b) as well. Necessary condition (c) involves restrictions that are imposed on any act and method to be used if it has or may have an impact on persons' liberties. Again, if the human operator and the computer are indistinguishable from each other in respect of activity in the problem area, it would be reasonable to assume that they employ equivalent sets of restrictions, so that necessary condition (c) is also satisfied. Hence, if the human operator is adequate in terms of showing the required respect to human dignity and the computer is indistinguishable from the operator, as far as activity in the problem area is concerned, the computer is also adequate in the same respect and to the same extent, and there is, therefore, no reason to avoid using it as a source of decisions and actions that ought to show respect for human dignity.

At this point, adherents of the Human Dignity Argument might raise an objection. So far, what has been shown is that a computer can be imagined that would be indistinguishable from a human operator in showing respect to human dignity. However, the very fact that the agent is not a person is incompatible with showing respect to the human dignity of targeted persons. Using the notion of "zombies", beings who are physically and functionally exactly like human beings but who are without any mental life, the objection is that it would be immoral, in the sense of being incompatible with respect of human dignity, to allow a zombie to kill persons.

For our rejoinder, consider first a micro or nano robot to be used for medical purposes. It is being given the coordinates of a tumor within a human body and additional data about that body and then properly inserted into the body. It navigates within the body, reaches the tumor, removes it, navigates back, and leaves the body. When it is regularly used in performing successful operations, will we be justified in ascribing it the responsibility for saving human life? Although this is the way many people are going to portray the robot, strictly speaking it won't be accurate to ascribe it responsibility. It is not the robot who shoulders responsibility but rather its operator, even if the latter does not have constant control over the activity of the robot. There are persons behind the robot in the field of responsibility around it.

Consider now a mine that explodes under the feet of a person or a bullet that hits his head. Is the mine or the bullet responsible for the death or injury of the person, or is it rather the person who planted the mine or shot the bullet? The killing drone is not essentially different from a mine or a bullet, even though it is incomparably more sophisticated. There is always a person who shoulders responsibility, and it does not matter whether he employs a simple mine or gun or a sophisticated computer on a drone. Respect for human dignity does not take an essentially different form when we move from mines and bullets to drones and their weaponry. It ought to be shown in the operating person's intentions, reasoning, and decisions, not in the nature of the artefact used.

Another argument against the usage of killing drones rests on the observation that, unlike the human operator, the computer under consideration does not have emotions and therefore cannot make decisions that rest on emotional intelligence. Let us call it the Emotional Intelligence Argument. A possible rejoinder to such an argument will indulge in debates about the very possibility of ascribing emotions to computers. Philosophical theories about the nature of emotions and their manifestations will play a major role in these debates, but for the present purposes we do not have to take sides in the debates. Given that the human operator and the computer are indistinguishable from each other in practice, we may plausibly assume that if some form of emotional intelligence is involved in the activity of the human operator, then some equivalent capacity is embodied in the computer.[6]

Again, if an adherent of the Emotional Intelligence Argument is going to raise the objection that respect for human dignity requires that the operator responsible for causing the deaths of persons be equipped with emotional intelligence, our answer would be similar to the one we have just used with respect to the Human Dignity Argument. The system that causes death includes not only the drone but also some persons, who have made decisions to employ it under certain circumstances, using their emotional intelligence to the extent required.

6. The Responsibility Argument

We turn now to a brief discussion of the Responsibility Argument. The argument is seemingly simple: if a person is killed because he or she was shot, then responsibility for one's death should rest with whoever shot the person and caused the death. Computers do not shoulder responsibility for their actions; hence, in case of an autonomous killing drone that shot a person to death, no responsibility for the death is shouldered, which is unacceptable. Cases of unnatural death should involve an ascription of responsibility for the event. Similarly, accountability is required for wrong decisions.[7]

To see why the Responsibility Argument is not sound, let us replace expressions of "responsibility for" by expressions of "contribution to the occurrence of". This is indeed part of the meaning of "responsibility" when it is ascribed as a matter of fact, under some circumstances, and not as a matter of norm. When "responsibility" is used on the grounds of some norm, the expression "responsibility for" is related to "duties to contribute to the occurrence of". As a matter of fact, does the responsibility for a person's death rest with whoever shot the person to death? Often yes but not solely so. Additional agents are often involved in contributing to the occurrence of the event, for example a commander who ordered the killing soldier to shot the person to death. Similarly, when a computer contributes to the occurrence of a certain event, it is never the computer solely that does it. There are always persons who write programmes, check them, approve them for usage and who thereby contributed to the occurrence of that event. Hence, even if we should avoid ascribing responsibility to computers, it does not mean that the event of a person having been shot to death by an autonomous killing drone involves no ascription of responsibility for the occurrence of the fatal event to certain persons.

This argument is applicable not only to the simplest case of a fixed programme that runs in a computer or a smartphone but also to the highly sophisticated programmes that collect data about their context and draw lessons from their own experience. The latter can make surprising moves; in science fiction stories, they sometimes turn against their operators, creators, or humanity in general. However, whenever a computer acts in what is superficially described as an "unpredictable manner", the capacity to make such moves has been or should have been predictable, and therefore responsibility has not been removed from the shoulders of the persons who developed the programme, approved it for usage, and employed it under some circumstances.

Should one avoid ascribing responsibility to computers? Daily parlance seems to have taken for granted a negative answer. People ascribe responsibilities and other human traits to artefacts simpler than computers, such as (old, computer-free) cars and guns, as well as their components. Standard depictions of smartphone applications, such as navigation tools, appear to present applications in terms of human agency. However, previous century philosophy has taught us to be wary of "systematically misleading expressions".[8] A philosophical discussion of such expressions will show us that they rest on utterly wrong assumptions and create utterly misleading impressions. The depiction of computers in terms of human agency, which involves mental capacity and moral agency, is a conspicuous example.

Here we can use arguments put forward by Searle, such as his Chinese Room Argument, against what he dubbed "Strong Artificial Intelligence", the view that if an entity runs a successful computational programme, then it has mental capacity. Such a capacity is a necessary precondition of moral agency in a fully fledged sense of the notion that involves an ability to produce intentions and make reasonable attempts to fulfil them and an ability to posit norms and deliberately follow them.[9] Searle's arguments show that a programme can be fully successful in performing some task of a human nature without having the mental capacity that human beings have and use when they perform the same task. The first example was that of a programme that successfully answers in Chinese questions posed to it in Chinese without understanding Chinese at all.[10]

Talk of computers is, strictly speaking, talk of devices that are interpreted by their inventors and users as performing certain tasks. A chess-playing computer is viewed as such only on grounds of our interpretation of what it does. Without our interpretation, it would be meaningless to ascribe chess playing to it. Similarly, a navigation application shows us the way just because we interpret its activity to be navigational. The voice that tells us to turn right is a sophisticated manifestation of that interpretation.

A killing drone will include a programme and a weapon. The programme will have a causal capacity to operate the weapon and accordingly will have a causal capacity to cause the death of a person. However, the programme will do it without having any mental capacity, let alone a moral one. It would be meaningless to ascribe it responsibility. However, there will always be a person to shoulder responsibility for causing that death, namely the person who enabled the programme to operate the weapon under certain circumstances, when using the killing drone in combat. When artefacts are under consideration, the fields of responsibility are never empty.[11]

7. The Practical Relevance of the Imitation Game

We take our discussion of the arguments against the usage of an autonomous killing drone that has successfully passed the test of the Modular Generalized Imitation Game to have shown that there is no reason to avoid using a computer for certain purposes after it has been compellingly shown that the computer is indistinguishable from a human operator whose activity is adequate in solving problems in the problem area.

Turing's Imitation Game was introduced as a theoretical framework for philosophical discussions of computers and their abilities as compared to human cognitive and other mental abilities. However, interest in Modular Generalized Imitation Game could be of a philosophical nature, of which the previous sections are a modest example, but could also be of a practical nature. There have been numerous attempts to develop programmes that can be used in replacing human beings with computers in a variety of circumstances that involve risk. Each of these attempts involves an application of a Modular Generalized Imitation Game, possibly in a tacit way, to the programme under development. The question of how close we are in being successful in playing a certain Modular Generalized Imitation Game for a certain problem area is, therefore, often of practical interest.

As it happens, a while ago, in June 2014, the University of Reading announced that the Turing Test "was passed for the very first time by computer programme Eugene Gootsman during *Turing Test (2014)* held at the renowned Royal Society in London . . . 'Eugene' simulated a 13 year old boy and was developed in Saint Petersburg, Russia . . . Eugene managed to convince 33% of the human judges (30 judges took part. . .) that it was human".[12]

Notice that the conditions we set earlier for passing a Modular Generalized Imitation Game are different from those involved in the Turing Test. Whereas the latter is passed if 30% consider the programme to be indistinguishable from a person, on grounds of 5-minute rounds of the game, we require 100% success in rounds as long as is required for establishing indistinguishability. Our proposal involves the maximal threshold.

8. Modular morality

A computer that is indistinguishable from a human operator in solving problems in the area of using drones in combat acts in a way that manifests compliance with principles of moral significance such as the principles of distinction and proportionality as required by the just war tradition and implemented in part in Humanitarian International Law.

The nature of such a computer's moral capacity should be clarified. In a nutshell, there are two ways of producing such a moral capacity. Let us call them the "universal method" and the "modular method". The *universal method* requires the creation of a programme that is able to make moral decisions under numerous significantly different circumstances.[13] When usage of drones in combat activity is considered, the programme will be applied to the combat circumstances under consideration and will make the required moral decisions. The *modular method* involves a programme that is able to make moral decisions only within a certain problem area. This is why we call it "modular".

There is an interesting difference between the universal method and the modular one with respect to moral reliability. When we compare a programme that makes decisions concerning the usage of drones in combat, on the grounds of the modular method, to a human operator who makes decisions in the same problem area, we can rely on the encountered indistinguishability if it is manifest in an appropriately long and heterogeneous series of rounds of the Modular Generalized Imitation Game in the problem area of usage of drones in combat. On the other hand, when we compare a programme that makes decisions concerning usage of drones in combat, on the grounds of the universal method, to a human operator who makes decisions in that problem area, we would be reluctant to rely on the indistinguishability with respect to the problem under consideration if it is accompanied by clear distinguishability with respect to other problem areas. The latter effect will be taken to cast a shadow on the moral reliability of the programme, though it pertains to other problem areas, if the human operator is evaluated to make morally better decisions in these areas.

Such a difference between the application of a universal method and the application of a modular method to the same problem area is not a shallow difference between a narrow perspective that ignores other applications of morality and a broad perspective that attempts to take into account all applications of morality. The difference reflects two views about morality, with respect to the relationships between the most fundamental moral principles and the most practical moral decisions. According to a commonly held view, one does not need more than fundamental principles and sound arguments in order to reach a practical decision under any circumstances. We beg to differ. We take the fundamental components of morality, such as the duty to respect human dignity under all circumstances, to be generic in nature, that is to say, to require supplementary doctrines that render the principles applicable to practical situations. There is no way to directly derive, for example, the principle of distinction of just war tradition from the fundamental principle of not treating persons as merely means rather than as ends. Similarly, there is no way to explain the difference between tolerating some collateral damage when fighting enemies and tolerating no collateral damage when fighting domestic criminals without the appropriate introduction of doctrines that manifest the difference between military ethics and police ethics. A full defense of our modular approach to morality is indeed beyond the scope of the present chapter. Here suffice it to assume that our moral evaluation of the possibility of killing drones rests on using the Modular Generalized Imitation Game rather than a Universal Generalized Imitation Game.[14]

9. Conclusion

In conclusion, we would like to consider two additional aspects of the issue under consideration.

First, assuming that under development is a computer that has become quite similar though still distinguishable from an adequate human operator of drones, how similar to the performance of an adequately performing human operator should the performance of the computer be for its usage in real combat to be morally justifiable? A precondition of such usage is the military necessity of replacing human adequate operators by computers that are not as adequate. Usually, operators are available, and they do not face significant risks during their combat activity, and, if so, why replace them and diminish the moral level of performance? However, one can imagine some circumstances in which it would be reasonable, for reasons of military necessity, to consider the possibility of replacing some operators by computers. How reliable should such a computer be for it to be morally justifiable to use it at all?

To answer this question, we use the analogous case of putting to use inaccurate weaponry. Ideally, combat weaponry should accurately deliver explosives to the target or its vicinity. When ideal conditions do not obtain and the usage of the weaponry does not risk the troops who operate it, then the crucial consideration would be that of proportionality: does the military advantage to be gained by using the inaccurate weaponry justify the expected collateral damage to be caused by it. Commanders will apply the appropriate proportionality considerations in this case the way they do it in other cases. The same type of consideration applies to the case of using computers, even if they are imperfect, in performing tasks of a certain kind.

Second, we would like to briefly discuss a slippery slope argument. According to this argument, it would be advisable to presently ban the usage of drones for targeted killing performed by human operators because this usage is bound to lead to the future usage of killing drones.[15] The logic of a slippery slope argument is clear and well known. At stage 1, you do something that is generally regarded as proper. At stage 2, you do something quite similar but still different from what you did at stage 1. Then, in a series of stages, what you do in one of them is very similar but still different from what you did at the previous stage. After sufficiently many steps along the series, you reach a stage in which what you do is very different from what you did at stage 1 and is generally regarded as improper. In order to avoid reaching that stage, you should avoid doing what you allowed yourself to do at stage 1. Thus, if you do not want to find yourself using autonomous killing drones, don't use drones for targeted killing performed by human operators either.

Although arguments of that structure abound with respect to a whole variety of issues, they are all invalid. Strictly speaking, there is a hidden assumption that users of such arguments commonly make without including it in the presentations of their arguments. It is the assumption that once you start moving along the series of stages, you are bound to reach the final stage in which what you do is improper. Quite often this is a wrong assumption. On many occasions, we are not bound to reach the final stage, if we move from stage 1 to the second one. We are not bound to reach a stage in which we behead our rivals if we use our knife not only for cutting fruits but also for preparing a Thanksgiving turkey. Our cultures provide us with fences that function on seemingly slippery slopes that allow us to act as we wish at stage 1 and move along the series of stages without reaching the bottom of the slope that involves improper activity. To show that a certain slope is actually slippery, one ought to show that no fences function on it and no barriers can be added to it in order to effectively stop slipping on it until its undesirable bottom. No demonstration of such a lack of extant fences and possible barriers has been presented on the apparent slope between the usage in combat of drones operated by human beings and the usage of autonomous killing drones. Actually, such a demonstration is not forthcoming if our arguments have convinced the readers of this chapter. Our suggested Modular Generalized Imitation Game is a starting point for the creation of the required barrier between morally justifiable and morally unjustifiable usages of drones in combat.

Notes

1 See, for example, ICRAC (2013) and its criticism in Waser (2013).
2 For an excellent survey, see Oppy and Dowe (2011).
3 Strictly speaking, a computer that runs a certain programme is under discussion, but for the sake of brevity we'll use the terms "computer" and "programme" to refer to a computer programmed in a certain way.
4 It will immediately become clear why we call it "generalized". In a later section, it will become clear why we call it "modular".
5 The suggestion has been made that computers should exceed human moral judgement, but adopting it involves issues that are beyond the scope of our present discussion. See Allen, Varner, and Zinser (2000) and Wallach and Allen (2012).
6 For a preliminary related discussion, see Wallach and Allen (2012).
7 See, for example, O'Connell (2014).
8 Ryle (1932). For a recent discussion, see Julia Tanney (2015).
9 Searle (1981); Searle (1997). See also Searle (2004) and Moural (2003).
10 Roff (2013) maintains that in our times, "a mere instrument of war is now a soldier". On the grounds of Searle's arguments, we would argue that a killing drone is not a soldier, since it is a programme with no mental capacities and moral agency.
11 For a related discussion, see, for example, Asaro (2008). Sparrow (2007) argued that since the activities of a killing drone are unpredictable by the commanding operator, the commander should not be held responsible for causing unwarranted death because he did not control the drone. We don't share such a narrow meaning of control. If a commander send a soldier to kill people in a certain area, surprising results should not exempt the commander from responsibility. Control does not have to mean direct involvement in the details.
12 See http://www.reading.ac.uk/news-and-events/releases/PR583836.aspx. For an example of a conversation with Eugene Goostman, see http://time.com/2847900/eugene-goostman-turing-test/. Doubts have been cast on several components of the success story. See, for example, http://www.thestar.com/news/world/2014/06/13/was_the_turing_test_passed_not_everyone_thinks_so.html
13 For different approaches to the universal method, see Allen, Varner, and Zinser (2000); Asaro (2008); Wallach and Allen (2009); Wallach (2010); Gerdes and Ohrstrom (2013).

14 For a related discussion, see Wallach and Allen (2012). Notice that our interest in computers that are indistinguishable from human beings in performing tasks of certain kinds does not commit us to a general consequentialist approach to morality. Our discussion and conclusions are compatible with a deontological approach to morality.

15 For a more general slippery slope argument related to drones, see Ryan (2014). For an informed discussion of future developments, see Dahm (2015).

References

Allen, C, Varner, G, and Zinser, J. 2000. Prolegomena to Any Future Artificial Moral Agent. *Journal of Experimental and Theoretical Artificial Intelligence* 12: 251–261.

Asaro, PM. 2008. How Just Could a Robot War Be? in PE Brey, A Briggle, and K Waelbers (eds), *Current Issues in Computing and Philosophy*. IOS Press, 50–64.

Bergen, PL, and Rothenberg, D (eds). 2015. *Drone Wars, Transforming Conflict, Law and Policy*. Cambridge University Press.

Brey, PE, Briggle, A, and Waelbers, K (eds). 2008. *Current Issues in Computing and Philosophy*. IOS Press.

Dahm, WJA. 2015. "Drones" Now and What to Expect over the Next Ten Years, in PL Bergen and D Rothenberg (eds), *Drone Wars, Transforming Conflict, Law and Policy*. Cambridge University Press, 348–358.

Evangelista, M, and Shue, H (eds). 2014. *The American Way of Bombing, Changing Ethical and Legal Norms, From Flying Fortresses to Drones*. Cornell University Press.

Gerdes, A, and Ohrstrom, P. 2013. Preliminary Reflections on a Moral Turing Test. *ETHICOMP* 2013: 167–174.

ICRAC (International Committee for Robot Arms Control). 2013. *The Scientists' Call*. Available online at http://icrac.net/call/.

Moural, J. 2003. The Chinese Room Argument, in B Smith (ed), *John Searle*. Cambridge University Press, 214–260.

O'Connell, ME. 2014. Banning Autonomous Killing, in M Evangelista and H Shue (eds). *The American Way of Bombing, Changing Ethical and Legal Norms, From Flying Fortresses to Drones*. Cornell University Press, 224–236.

Oppy, G, and Dowe, D. 2011. The Turing Test. *Stanford Encyclopedia of Philosophy*. Available online at http://plato.stanford.edu/entries/turing-test/.

Roff, HM. 2013. Responsibility, Liability, and Lethal Autonomous Robots, in F Allhoff, NG Evans, and A Henschke (eds), *Routledge Handbook of Ethics and War: Just War Theory in the 21st Century*. Routledge, 352–364.

Ryan, K. 2014. What's Wrong with Drones? in M Evangelista and H Shue (eds), *The American Way of Bombing, Changing Ethical and Legal Norms, From Flying Fortresses to Drones*. Cornell University Press, 207–223.

Ryle, G. 1932. *Systematically Misleading Expressions*. Proceedings of the Aristotelian Society XXXII. Available online at www.ditext.com/ryle/sme.html.

Searle, JR. 1981. Minds, Brains, and Programs. *Behavioral and Brain Sciences* 3: 417–457.

———. 1997. *The Mystery of Consciousness*. New York Review of Books.

———. 2004. *Mind, a Brief Introduction*. Oxford University Press.

Smith, B (ed). 2003. *John Searle*. Cambridge University Press.

Sparrow, R. 2007. Killer Robots. *Journal of Applied Philosophy* 24: 62–77.

Tanney, J. 2015. Gilbert Ryle. *Stanford Encyclopedia of Philosophy.* Available online at Plato.stanford.edu/entries/ryle/.

Turing, A. 1950. Computing Machinery and Intelligence. *Mind* 59(236): 433–460.

Wallach, W. 2010. Robot Minds and Human Ethics: The Need of a Comprehensive Model of Moral Decision Making. *Ethics and Information Technology* 12: 243–250.
Wallach, W, and Allen, C. 2009. *Moral Machines: Teaching Robots Right from Wrong*. Oxford University Press.
———. 2012. *Hard Problems: Framing the Chinese Room in Which a Robot Takes a Moral Turing Test*. AISB/IACAP, University of Birmingham, 5 July 2012. Available online at https://www.researchgate.net/publication/289173777_Hard_problems_Framing_the_Chinese_room_in_which_a_robot_takes_a_moral_turing_test.
Waser, MR. 2013. *The Bright Red Line of Responsibility*. IACAP Proceedings. Available online at www.IACAP.org/proceedings_IACAP13/paper_29.pdf.

7 Delegation and Responsibility

A Human–Machine Perspective

Tjerk de Greef

1. Introduction[1]

This following chapter discusses the concept of moral responsibility from a sociotechnical perspective, which is known in other fields as Human Factors or Cognitive Ergonomics. A key issue within autonomous drones relates to the responsibility gap, and this chapter claims that non-moral agents can't be held responsible, bringing the concept of command responsibility to the foreground. We need to look at delegation rather then autonomy. Furthermore, the phrase 'autonomy' is used differently across disciplines, leading to ambiguity as to what it means to be autonomous. Therefore, a distinction between moral and operational autonomy is important and explains the difference between human and artificial actors. This chapter focuses on a human–machine agent perspective, discussing delegation and responsibility from an engineering perspective.

Drones are considered strategic assets for surveillance and reconnaissance tasks. Significant investments have increased the possibility of controlling such technology from great distances, thereby rendering boots on the ground unnecessary. Clearly, remotely piloted vehicles are on a path of higher levels of autonomy, and it seems a matter of time that at least parts of the missions will be flown autonomously. Higher levels of autonomy require deep insight into human–machine interaction paradigms. Optimizing the way humans and intelligent technologies interact has been a research topic for decades. But in order to reach an optimal human–machine collaboration, a vast repository of research is needed to attain the right balance between technologically feasible levels of autonomy and human requirements and limitations. In other words, interaction with highly technological systems requires a seamless integration of the human and the machine to ensure a proper fit of human capabilities in the appropriate context, to avoid misuse, and to prevent errors and accidents. In this volume's next chapter, van Wynsberghe and Nagenborg acknowledge that design is a valid starting point to 'humanize' the use of highly advanced technologies. The last decade is marked by a shift towards '*partnering with*' rather than '*automation extending*' human functioning. The ultimate goal is to create a marriage between the human and the computerized entity in order to act in symbiosis, creating a powerful cooperation that extends the sum of the single entities because neither is able to solve problems individually.

Furthermore, a debate on a responsibility gap emerges as a result from these increased levels of autonomy. The responsibility gap focuses on the question of whom to hold responsible in case of misdoing. But the discussion on the responsibility gap is far subtler. Even when a drone operates a mission fully autonomously, someone within the chain of command has made the decision to deploy the autonomous drone for a particular operation and thereby has responsibility. Now the step to command responsibility is easily made. Such an organizational perspective –including questions of responsibility – is completely missing in the responsibility gap discussions and ignores that there is always some sort of organizational structure in any team or group composition. Furthermore, when a task is delegated, the responsibility question rests on the case whether the delegate is considered being a moral agent or not. This is not the case with non-moral agents because non-moral agents may '*do*' things, but they cannot be held responsible for what they do. Not holding autonomous drones responsible is also claimed by Müller (Chapter 3 in this volume).

This chapter therefore proposes that the human is fully responsible for what the machine does once tasks are delegated to the machine. The human delegating work to the machine has a moral obligation to supervise the computer in its activities (what it being done for which reasons). To enhance the supervision process, this chapter also proposes an *accessible transparency* criterion.

This chapter is organized as follows. Section 2 puts drones in context, including contemporary and future perspectives. Section 3 discusses distance vs. autonomy and treats both as independent. Section 4 introduces a joint cognitive systems perspective to ensure a proper collaboration among drones that are partially autonomous. Section 5 discusses delegation and responsibility in light of the proposed joint cognitive systems architecture. Prior to conclusions, Section 6 proposes the criterion of accessible transparency, allowing supervising a partial autonomous machine in an easy and natural fashion.

2. Military Drones in Context

Large investments in military technology, including phased array radars, missile guidance systems, and military robots, have changed the character of war. In this context, Michael Ignatieff is often quoted when commenting on NATO's intervention in Kosovo by stating that soldiers "*were removed from the actual combat zones, but carried out military missions with the help of technology*" (Leveringhaus & de Greef 2013). Ignatieff wrote this in 1999. Since then, investments in drone technologies further reduced the need to have boots on the ground. Drones are typically launched in the vicinity of an area of interest but controlled from a remote control cabinet using a satellite relay system. Although deploying drones is controversial for the practice of targeted killings and signature strikes, they are considered a strategic asset for surveillance and reconnaissance tasks. Drones, whose name is based on the constant, monotone, and humming sound that arises from the propeller engines, are also known as unmanned aerial vehicles. Military organizations prefer to refer to such unmanned systems as remotely piloted vehicles to emphasize that these vehicles are actually operated by well-trained pilots, although not from the in situ cockpit but from a different location typically thousands of miles away.

Clearly, unmanned or remotely piloted vehicles are, from a technology perspective, on the path of higher levels of autonomy, and it seems a matter of time, notwithstanding the campaign against the so-called Killer Robot (Human Rights Watch 2012) and the moratorium on autonomous systems by the United Nations (Heyns 2013), that at least parts of the missions will be flown autonomously. And the emergence of higher levels of autonomy is not merely the result of technology that is ready for deployment but also the insight that it is a way to reduce the degree of man–machine interactions, thereby lowering the cost of ownership. The Tanaris drone, for example, has – in addition to using a set of stealth technologies to reduce radar reflections and emissions – capabilities to act semi-autonomously, although its publicly available specifications lack a clear definition of what 'semi-autonomous' exactly entails. The website of BEA Systems offers a leaflet describing a future field test where the Tanaris would search an area of interest using a preprogrammed flight path to locate and identify a target upon seeking acknowledgement from mission control to neutralize the target (BEASystems 2011). This description would fit a category where the human would still be 'on the loop' and Tanaris would be flying parts of the mission without human input. On the other hand, the X47-B drone has shown the capability of taking off and landing without any human intervention on sailing aircraft carriers. Also, the X47-B is capable of refuelling without human intervention. In this case, it seems as though the X47-B acts autonomously more for landing and refuelling than for other tasks.

As stated, drones are considered very useful in surveillance and reconnaissance tasks. These types of tasks are different compared to robot technologies that execute the *dull, dangerous, and dirty* work. Examples are robots used for mine or bomb disposal purposes, deep sea missions, urban combat reconnaissance (e.g. the Dragon Runner), load carrying (e.g. the Alpha Dog), and powered exoskeletons (e.g. HULC). In these, the high-level goal is to have human operators remain effective while not experiencing cognitive underload or extreme levels of stress resulting from activities that directly endanger their lives.

3. Distance vs. Autonomy

When discussing the future of drones, it is important to distinguish between effects due to physical distance and effects resulting from higher levels of autonomy. In the context of drones and physical distance, one should recognize that operators experience the environment through on-board sensors that affect the way operators comprehend the battlefield. This has benefits as well as downsides for decision-making and showcases the need for properly designing human–computer interaction systems. And with regard to increased autonomy, there is a growing concern that ethical values are violated up to the point that we are losing humanity (Human Rights Watch 2012; Sparrow 2007; Matthias 2004). Both are thus important factors and should be regarded as independent.[2] The following section elaborates on these effects as a result of either distance or increased levels of autonomy.

3.1 Distance to the Battlefield

Artillery guns are considered the first weapon systems that increased the distance of participants from the battlefield. Ever since, ethicists have debated the effect of distance on decision-making. Clearly, drones showcase another technology that increases the physical distance to the battlefield. Typically, the distance between the drone and the control cabinet is easily a couple of thousand miles. As a result of these large distances, many worry that operators get morally disengaged (Royakkers & van Est 2010). A larger physical distance to the battlefield has benefits as well as downsides. First and fairly obvious, the large distance reduces the risks for operators themselves.[3] Second, given that they are not within the combat zones, the experienced stress is diminished. It is well known that stress affects decision-making because it influences how human beings interpret their environment and frame certain issues (Porcelli & Delgado 2009; Keinan 1987). A decrease in stress reduces tunnel vision and leads to greater situational awareness, as well as more accurate interpretations of morally relevant facts. Following these arguments, the effect of increased distance on decision-making is rather positive. But the opposite is also true. Boredom or underload negatively affects decision-making. Endsley and Kiris (1995) showcased that being too much out of the loop reduces the quality of decision-making and leads to performance degradation. It is thus important to find the sweet spot between distance and stressors in order to optimize decision-making, tunnel vision, and performance.

Drones, or remotely piloted vehicles, have two distinct characteristics compared to the piloted jet fighters. First, drones have *increased surveillance capabilities* due a rich sensor suite. Such a modern sensor suite entails not only more sensors but also better-quality sensors. Although more information is considered better, it also raises issues of information overload. Also, drones have a *larger surveillance capacity* due to increased flying time compared to jet fighters or helicopters. Both characteristics lead, at least theoretically, to an improved recognition of *patterns of life*, thereby reducing, in turn, collateral damage. In other words, improved situational assessment allows timing any attack such that the expected collateral damage is the least it can be.

In the context of drones, a number of downsides are also reported. First, drones operators have an increased risk of psychological injury, which some refer to as moral injury (Gregory 2011), such as post-traumatic stress disorders (PTSD) due to the fact that some drone pilots bond with objects within the area of interest (Abé 2012). Second, soldiers are directly removed from the horrors of war, and there is the risk that operators see the enemy not as humans but as a blip on a screen. This is, by some, referred to as the game mentality or moral disengagement (Sharkey 2010; Royakkers & van Est 2010). Third, it is worried that operators become trigger-happy with remote-controlled armaments, situated as they are in complete safety, far from the conflict zone.

The distance between the operator and war zone is now basically maximized and has affected how operators perceive the environment, also affecting the comprehension of morally relevant facts. The effects can be positive for decision-making, but there are also downsides, and they showcase the need for properly designing human–computer interaction systems.

3.2 Autonomy

Many military organizations consider equipping drones with high levels of autonomy in order to reduce the degree and amount of interaction between the machine and its operators, thereby lowering the operational cost. However, some argue that ethical values come under pressure as a result of high levels of autonomy, up to the point that a critical threshold is passed (Sharkey 2010; Royakkers & van Est 2010; Cummings 2003). Placing the human operator outside the decision-making loop, Human Right Watch (2012) argues, opens up a responsibility gap – that is an ambiguity with respect to whom should be held responsible in case of misdoing (Sparrow 2007; Matthias 2004). Is it the robot, the programmer, or the army?

But I believe that the discussion on the responsibility gap is far subtler for two reasons. First, when a drone operates a mission in full autonomous mode – that is that the drone achieves a military objective without any human input or control – someone within the chain of command has made the decision to deploy the autonomous drone for the particular operation and thereby has responsibility. One could look at it from a hierarchical perspective because command responsibility has existed since 1899. More broadly defined, it refers to the duty to supervise subordinates and claims liability for the failure to do so. Such a perspective is completely lacking the debate on full autonomous systems and ignores the facts that there is always an organizational structure.

Second, the notion of autonomy seems to be used differently across disciplines, leading to ambiguity as to what it means to be autonomous. Philosophers fall back on Immanuel Kant (1724–1804) when defining moral autonomy. In this sense, autonomy is defined as a capability to (a) pursue the concept of good life (cf. express free will) and (b) develop desires about our desires (cf. second-order desires). The latter allows humans to distance ourselves from our initial beliefs and critically reflect upon them. By contrast, the discipline of human factors interprets autonomy in the context of human–machine interaction, which we will refer to as operational autonomy.

Following our definition of operational autonomy, Bradshaw and colleagues (2004) claim that autonomy has two senses: self-sufficiency and self-directedness. The first, self-sufficiency, relates to the set of actions that are possible with or without the help of another actor whereas the second, self-directedness, relates to the set of actions that are permitted or required to be executed.[4] Low self-directedness indicates that, although potentially capable of performing the task, the agent is not permitted to do so, whereas high self-directedness indicates that the agent has authority over its own actions, though it does not necessarily imply sufficient competence. For instance, rules of engagement provide explicit criteria for the use of force as well as for the utilisation of certain components of weapons systems. A pilot of a fighter jet might be constrained to use its fire radar sensor to lock on suspected air tracks. The reason for this prohibition is that opposing forces could interpret this as a hostile intention, possibly triggering the use of force. Likewise, a drone might be restricted in its use of specific cameras due to privacy regulations.

The self-sufficiency axis of autonomy matches nicely with the level of automation taxonomies proposed in the field of human factors. Various studies have been conducted that provide indications on the level of control that can be allocated towards a human or a machine. These studies aim to find the sweet spot between full autonomy and manual operations given the use of the technology in the context of deployment. Such intermediate levels, often denoted as mixed initiative, shared control, or adaptive/adaptable automation, are seen as the most promising area for improved efficiency in interactive systems (de Greef et al. 2007).

3.3 Operational Autonomy in a Team

The last decade has been marked by a shift towards '*partnering with*' rather than '*automation extending*' human functioning (de Greef 2012), and such approaches are annotated as quasi team members, junior crew members, ePartners, or agents as part of a human–agent team. The ultimate goal is to create a marriage between the human and computerized entity to act in symbiosis, creating a powerful cooperation that extends the sum of the single entities because neither is able to solve problems individually. The well cited paper by Klein and colleagues (Klein et al. 2004) describes 10 challenges to make automation a team player and closely matches the joint cognitive systems approach (Hollnagel & Woods 2005). When designing a joint cognitive system, a clear focus has to be on the joint activity while operating within safety boundaries (cf. Salas et al. 2008: 544), which addresses nicely the self-sufficiency sense (cf. joint activity) and the self-directedness sense (cf. within boundaries).

4. Towards Joint Cognitive Systems

In 1951, Paul Fitts published a list that supported designers to allocate tasks to either the machine or the human operator based upon intrinsic capabilities. In the decades that followed, alternative man–machine human computer interaction (HCI) paradigms emerged when Fitts's *all-or-nothing* assumption proved poor. Roughly, three generations of HCI paradigm are recognized. The first generation is based upon the seminal work of Sheridan & Verplank (1978), who defined 10 Levels of Automation (LoA) that differentiate executed actions and information offered by the machine. The second generation of HCI systems were heavily founded upon human information-processing strategies and Parasuraman and colleagues (2000) claim that all of these stages should be automated at different levels of automation based on primary (e.g. human performance) and secondary (e.g. automation reliability) criteria. The third generation builds upon the work of Parasuraman and colleagues (2000) but focuses on the question of how to divide the work within each of the four stages (Arciszewski et al. 2009). The latter model recognizes five different levels of automation that are clearly differentiated by (a) whether the machine or the human engages in action and (b) whether the machine actively signals the human. Furthermore, the Arciszewski model focuses on the central domain objects and their attributes, easily leading to a fine-grained and natural division of labour between the human and machine. Working agreements, which allow making explicit which tasks and objects are in control of the human or the machine, facilitates such a fine-grained division of labour.

Figure 7.1 proposes an architecture for a joint cognitive system and reveals two distinctive features (based upon Arciszewski et al. 2009): the concurrency feature and the working agreements feature. Both are keystone aspects towards natural interaction. The concurrency feature acknowledges that the machine and the human might have different interpretations requiring both entities to do their job concurrently. Secondly, working agreements allow the human to explicate a covenant about the division of work between the human and the machine.

First, the *concurrency* feature allows both parties do their job. Each entity arrives concurrently at its own *representation* of the world based upon their cognitive reasoning and algorithmic mechanisms. Each deposits their respective cognitive and computational efforts in their *storage space*, including the underlying information. Having these two views on the world is akin to two people coming to different conclusions based on the same set of (non-conclusive) data. The machine is thus seen as a team member with a limited but scrupulously objective view of the world.

Using the concurrency feature, five distinct levels of automation follow logically when (a) self-sufficiency is being shifted between the human and the machine and (b) whether signalling is executed actively by the machine or not (see Table 7.1). The *advice* level of automation represents the level where the machine world view is available for advice. In other words, the human can always consult the machine for its opinion, thereby serving as a *second opinion*. In the *consent* level

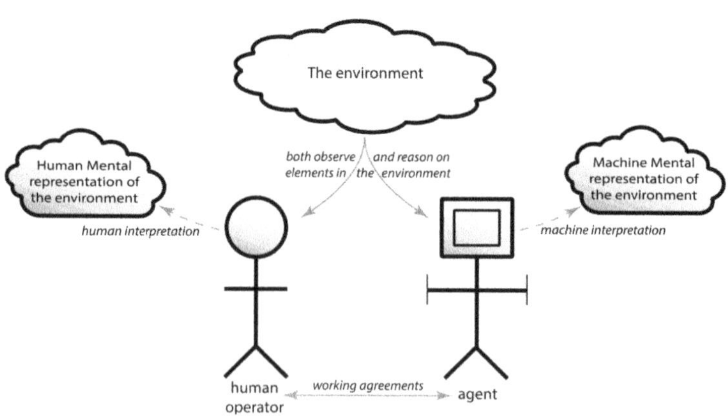

Figure 7.1 The proposed joint cognitive system

Table 7.1 Five levels of automation, as defined by Arciszewski et al. (2009), that are differentiated by whether the machine or the human engages in action and whether the machine actively signals the human

Level of Automation		*Description*	*Self-Sufficiency*	*Signalling*
5	System	The human delegates everything to the automation, and there is no interaction.	No (with machine)	No
4	Veto	The human delegates work to the automation *unless overruled* (vetoed).	No (with machine)	Yes
3	Consent	The human keeps all work to itself, but the machine alerts the human to changes in the situation.	Yes (with human)	Yes
2	Advice	The human keeps all the work to itself, but the automation is available for advice (but takes no initiative).	Yes (with human)	No
1	Manual	No automation is available.	Yes (with human)	No

of automation, the machine compares both views and actively signals any discrepancies to the human. The machine view is, of course, still available for inspection. Signalling occurs only when a discrepancy occurs or when additional evidence is found (by the machine) for a renewed machine world view. Whereas the advice level of the automation denotes passivity, whenever the human actively pulls the view of the machine, the level becomes active, and the machine actively pushes its opinion to the user. However, the user remains in control of the actual decision.

At the other end of the spectrum, the machine is granted the highest level of automation. At this *system* level of automation, the machine actively pushes its opinion to the human but without actual signalling. One could speak here in terms of full delegation of work as the machine does the work and doesn't bother the human with its decision. In contrast, at the *veto* level of automation, the machine is granted the same responsibility in terms of decision-making but also alerts the human about actions taken, thereby allowing the human to intervene. The only difference between these two levels is that in veto mode, the system warns the human of its decision, thereby giving the human the option to reject and 'veto' the decision of the machine, whereas in system mode the machine performs the same action in silence. The *manual* level of automation is added to denote the fact that no automation is present for whatever reason (technological, ergonomic, political, financial, safety).

Working agreements serve as explicit arrangements between the human and the machine about the division of work (Arciszewski et al. 2009; Miller & Parasuraman 2007; Parasuraman & Miller 2004). Working agreements are based upon predefined tasks (e.g. identification) and object attributes (e.g. height, velocity, air traffic lane). Table 7.2 shows a couple of examples, showcasing the naturalness of

Table 7.2 Two example of working agreement definitions for the task identification

Working Agreement	*Set Definition* (in words)	*Set Definition* (in logic and related to LoA)
1	Let the user identify all objects, limit the system to warning the user when something seems amiss with a track.	**For all** objects: *consent LoA*
2	Let the system identify all planes above 30,000 feet in air traffic corridors (ATC).	**IF** altitude > 30,000 feet **AND** within ATC: *system LoA* **ELSE** others: *consent LoA*

how agreements are formalized. In agreement 2, for example, the machine does more work as it processes those objects that fly high in the air and are flying in the air traffic corridor (ATC), whereas in agreement 1, the human does most of the decision-making. More complex agreements are easily facilitated using additional object attributes. Working agreements allow the human to arrange the division of work.

Working agreements minimize the automation–human coordination asymmetry (Woods et al. 2004) because working agreements define an a priori explicit arrangement *what* and *what not* to delegate to the automation. Working agreements can reflect easily the opinions of the system designers and human on the proficiency of the software and can also serve to establish trust in the workings of the machine by starting at fairly low levels of automation for the whole set of objects. They can also mirror thoughts on how much work is ultimately delegated to the machine under the political and strategic constraints of the mission.

5. Delegation in Light of Joint Cognitive Systems[5]

The responsibility discussion on (semi-)autonomous systems fails to recognize an appropriate understanding of task delegation and underexplores significant achievements in this area. For example, if I delegate work to you, does it mean that you are fully responsible? Or do I still have a responsibility alike command responsibility, which refers to the duty to supervise subordinates and introduces liability for the failure to do so? To clarify, a relationship between two highly trained professionals is totally different compared to a father–child relationship. This raises very interesting questions, and we need a proper understanding of the effect of delegation on responsibility.

The most important question is whether you delegate a task to another moral agent or to a non-moral agent. The difference between the two is that we can hold moral agents responsible for what they do – we can praise them and blame them (that is, develop justified reactive attitudes towards them). This is not the case with non-moral agents because non-moral agents may 'do' things, but they cannot be held responsible for what they do.

If you have delegation to a machine in mind, then this is delegation of work to a non-moral agent, which means that the machine cannot be held responsible for what it does. The human, having delegated the task to the machine, is fully responsible for what the machine does. This is because the human has delegated the task and instructed the machine on what to do. Following the concept of command responsibility, the human has an obligation to supervise the computer on its activities (what is being done and for which reasons). However, to enhance such an obligation and to facilitate transparency, the criterion of *accessible transparency* is now discussed.

6. The Criterion of Accessible Transparency

In the joint cognitive system approach, machine is considered a team player, and this section proposes that both the human and the artificial agent build their own interpretation of the world *concurrently*. Furthermore, working agreements allow dividing the work in a natural fashion. To allow for easy and natural supervision, the design principle of *accessible transparency* is proposed. Meeting the accessible transparency criterion assures a minimal effort to accessing data underlying the machine interpretation with a fingertip, significantly increasing transparency and minimizing automation surprises.

Yaniv and Kleinberger (2000) suggest that an advice by a machine (cf. machine interpretation of the world in Figure 7.1) is underused because humans fail to have access to the rationale underlying the advice. In other words, there exists a risk that the human doesn't appreciate the interpretation of the machine because the underlying data are not easy to access or/and comprehend. A common assumption in cognitive psychology is that the weight placed on a judgment depends on the evidence that is recruited to support that judgment (Tversky & Koehler 1994). Thus, because the argumentation underlying the interpretation of the machine is not available for cognitive reasoning, the importance of the machine interpretation becomes less useful than expected. In situations where people receive advice they did not expect, they ask questions like "*why*" and "*why not*" and seek an explanation. An explanation is used to infer the quality of the interpretation since it offers insight into how the advice was reached (Bonaccio & Dalal 2006). This includes revealing in a clear fashion that the machine was missing particular data. Such explanations increase trust when the human believes the right arguments are provided but may equally decrease trust when the arguments make no sense. Having access to the underlying data thus provides more ways to assess the quality of advice.

Imagine an interaction display for a military identification task, where both the human and the machine arrive at a particular identity (hostile, suspect, neutral, assumed friendly, friendly, or unknown) based an identification rule set. For each of the six identities, a unique button exists (organized in a row). The display expresses in a color representation the belief of each entity in a particular identity. As an example, consider that the human, represented in, say, blue, believes it is a neutral object and the machine, represented in yellow, believes it is a suspect object. This is expressed by coloring the appropriate buttons, respectively, blue and yellow. Now the human can press, for example, the suspect button, thus being the machine interpretation, and the display responds by highlighting the arguments in favor (in green) or against (in red) for the claim that the object has a suspect identity (according to the machine). Missing information is represented in dark yellow. Thus, a simple touch by the user shows which data conflict and which support the selected identity. In this particular case, the machine based its belief in a suspect identity on the fact that the object is *maneuvering* and has *IFF 4*. These two data fields will turn green. However, the selected object is in an *ATC* (air traffic corridor) that contradicts the suspect identity. The corresponding data field will turn red. Despite this conflicting information, the machine acknowledged the suspect identity. This is a good example of meeting the accessible transparency criteria, where arguments are easily visualized at a fingertip away.

Allowing the machine rationale a fingertip away allows the human to develop – in time – an understanding of machine behavior and beliefs. Eventually this will lead to the development of trust as it is better understood what the machine is capable of doing and under which circumstances. Making sure that underlying information is naturally available allows supervising the artificial agent in an easy fashion. In an experiment with eight navy officers (de Greef et al. 2010), it has been observed that some officers replied that they did not agree with the machine but that they did understand the machine. This not only showcases that human and machine disagree when offered the same information; it also shows that meeting the accessible transparency criterion allows for an improved comprehension of machine behavior and a minimization of automation surprises.

Conclusion: Towards Responsible Interaction

Although deploying drones is controversial for the practice of targeted killings and signature strikes, they are considered a strategic asset for surveillance and reconnaissance tasks. Large investments in drone technologies have further reduced the need to have boots on the ground. Clearly, unmanned or remotely piloted vehicles are, from a technology perspective, on the path of higher levels of autonomy, and it seems a matter of time that at least parts of the missions will be flown autonomously. And the emergence of higher levels of autonomy are not merely the result of technology that is ready for deployment but also the result of growing awareness that it is a means to reduce the degree of human machine interaction, thereby lowering the cost of ownership.

Interaction with highly technological systems requires a seamless integration of the human and the machine, ensuring a proper fit of human capabilities, omitting the misuse of the systems, and preventing errors and accidents. The last five decades proposed a variety of human–machine interaction models, including models of how to divide the work between entities. These insights accumulated into a concept where automation is considered a team player.

This chapter proposes a soft architecture that reveals two important features. First, instead of letting a task be performed either by the intelligent agent or the human, both entities create an interpretation of the world *concurrency*. In this way, both the human and the intelligent machine arrive at their own interpretations of the situation. Second, working agreements provide a natural basis to define a priori a division of work using five distinct levels of automation that deviate in the levels of signaling and authority.

Supervising intelligent systems, including future drones, is a necessity to avoid the responsibility gap. The concept of command responsibility is relevant here and refers to the duty to supervise subordinates and introduces liability for the failure to do so. Being responsible depends on whether a task is delegated to another moral agent or a non-moral agent. If work is delegated to a machine, then the human delegates a task to a non-moral agent, which means that the machine cannot be held responsible for what it does. The human, having delegated the task to the machine, is fully responsible for what the machine does. This is because that person delegated the task and instructed the machine on what to do.

To facilitate supervision, transparency becomes relevant, and the criterion of *accessible transparency* is proposed as a means to do so. Supervision requires understanding the activities of the computer (what it being done and for which reasons). Meeting the accessible transparency criterion assures a minimal effort to access data underlying the machine interpretation at a fingertip away. Allowing the machine rationale a fingertip away allows the human to develop – in time – an understanding of machine behavior and beliefs. Eventually this will lead to the development of trust because what the machine is capable of doing and in which circumstances are better understood. Making sure that underlying information is naturally available allows supervising the artificial agent. However, the effect of concepts like accessible transparency needs to be studied carefully in the context of decision-making and in particular with respect to the effects of moral injury (Abé 2012; Gregory 2011) or the game mentality (Sharkey 2010; Royakkers & van Est 2010).

Kill decisions also require attention. This chapter proposes an architecture that facilitates considering a highly automated agent as a junior team member (Figure 7.1). A junior team member can, dependent on prior experience in a particular context, come to the benefit of human decision-making, which can be at various stages in the decision-making process (e.g. identification or deployment of weapons) and at different levels of automation. But – and this can't be stress enough – the principle of command responsibility remains, making the one who ordered the activity ultimately responsible.

This chapter provides a human–machine perspective on future drones that are envisioned toward higher levels of autonomy. Adding delegation perspective clarifies that the human operator maintain responsibility in such systems, but it is important to further optimize human–machine interaction by designing proper interfaces between the human and intelligent systems that also explain the rationale. Taking the concept of delegation as a design principle is a stepping stone towards designing responsible systems.

Notes

1 A significant part the work was funded by the *Military Human Enhancement: Design for Responsibility in Combat Systems* project (No: 313-99-260) that was financed by the Netherlands Organization for Scientific Research (NWO) in the program line Societal Responsible Innovation. But the presented insights and models were also the result of the

Collaboration at a Distance project that was run in the IOP-MMI program by SenterNovem (an agency of the Netherlands Ministry of Economic Affairs) and the Smart Operations programme V929 that was hosted by the Netherlands Research and Defense organization under auspice of Netherlands Organization for Applied Scientific Research (TNO).

2 Independence between distance and autonomy is often not respected. A typical debate starts by addressing the effects of physical distance on decision-making, thereby yielding a yuck statement that high levels of autonomy are inherently bad. High levels of autonomy are in themselves not unproblematic but for different reasons.

3 Assuming that areas with control cabinets are not considered war zones; if so, they might also become a subject of attack. Notwithstanding, hitting these zones is significantly complicated as non–war zones surround these areas.

4 Ron Arkin prefers to talk about constraints instead of permitted and obliged actions as the laws of armed conflict constrain the deployment of force rather than oblige the use of it (Arkin 2009).

5 I would like to thank Dr Alexander Leveringhaus for our frequent and passionate discussions on these matters when we both were working on the design for responsibility in the Military Human Enhancement project. Combining the view from a philosopher and a computer scientist greatly helped to disambiguate matters that worked to the advantage of both disciplines.

References

Abé, N., 2012. Dreams in Infrared: The Woes of an American Drone Operator (Part 2: An Unpopular Job). *Der Spiegel*, December 14.

Arciszewski, H.F.R., de Greef, T.E., & van Delft, J.H. 2009. Adaptive automation in a Naval Combat Management System. *IEEE Transactions on Systems, Man, and Cybernetics Part A: Systems and Humans* 39(6): 1188–1199.

Arkin, R., 2009. *Governing Lethal Behaviour in Autonomous Robots*, Boca Raton, FL: Taylor & Francis.

BEASystems, 2011. *Taranis – Looking to the Future*. San Jose, CA: Author. Available online at http://www.baesystems.com/en/product/taranis (accessed 2 March 2016).

Bonaccio, S., & Dalal, R.S., 2006. Advice Taking and Decision-making: An Integrative Literature Review, and Implications for the Organizational Sciences. *Organizational Behavior and Human Decision Processes* 101(2): 127–151.

Bradshaw et al. 2004. Ten Challenges for Marking Automation a "Team Player" in Joint Human-Agent Activity. IEEE Intelligent Systems, 2004 19(6): 91–95.

Cummings, M.L., 2003. Automation and Accountability in Decision Support System Interface Design. *The Journal of Technology Studies* 32: 23–31.

De Greef, T., 2012. *ePartners for Dynamic Task Allocation and Coordination*. Dissertation, Delft University of Technology, Delft.

De Greef, T.E., Arciszewski, H.F.R., & Neerincx, M.A., 2010. Adaptive Automation Based on an Object-Oriented Task Model: Implementation and Evaluation in a Realistic C2 Environment. *Journal of Cognitive Engineering and Decision Making*, 31: 152–182.

De Greef, T.E., van Dongen, K., Grootjen, M., & Lindenburg, J., 2007. Augmenting Cognition: Reviewing the Symbiotic Relation between Man and Machine, in D.D. Schmorrow & L.M. Reeves (eds). *Proceedings of the 3rd International Conference of Foundation of Augmented Cognition held as part of HCI International 2007*. Berlin, Heidelberg: Springer Verlag, 439–448.

Endsley, M., & Kiris, E., 1995. The Out-of-the-loop Performance Problem and Level of Control in Automation. *Human Factors* 37(2): 381–394.

Fitts, P.M. (ed), 1951. *Human Engineering for an Effective Air Navigation and Traffic Control System*. Washington, DC: National Research Council.

Gregory, D., 2011. From a View to a Kill: Drones and Late Modern War. *Theory, Culture & Society* 28(7–8): 188–215.

Heyns, C., 2013. *Report of the Special Rapporteur on Extrajudicial, Summary or Arbitrary Executions.* UN General Assembly, Human Rights Council, A/HRC/23/47.

Hollnagel, E., & Woods, D.D., 2005. *Joint Cognitive Systems: Foundations of Cognitive Systems Engineering*, Boca Raton, FL: Taylor & Francis.

Human Rights Watch, 2012. *Losing Humanity: The Case against Killer Robots*. Available online at https://www.hrw.org/report/2012/11/19/losing-humanity/case-against-killer-robots (accessed 28 January 2016).

Keinan, G., 1987. Decision Making under Stress: Scanning of Alternatives under Controllable and Uncontrollable Threats. *Journal of Personality and Social Psychology* 52(3): 639–644.

Klien, G., Woods, D.D., Bradshaw, J.M., Hoffman, R.R., & Feltovich, P.J., 2004. Ten Challenges for Marking Automation a "Team Player" in Joint Human–Agent Activity. *IEEE Intelligent Systems* 19(6): 91–95.

Leveringhaus, A., & de Greef, T., 2013. Tele-operated Weapons Systems: Safeguarding Moral Perception and Responsibility, in M. Aaronson & A. Johnson (eds). *Hitting the Target: How New Capabilities Are Shaping International Intervention.* London: The Royal United Services Institute for Defence and Security Studies, 57–64.

Matthias, A., 2004. The Responsibility Gap. Ascribing Responsibility for the Actions of Learning Automata. *Ethics and Information Technology* (6): 175–183.

Miller, C.A., & Parasuraman, R., 2007. Designing for Flexible Interaction between Humans and Automation: Delegation Interfaces for Supervisory Control. *Human Factors* 49(1): 57–75.

Parasuraman, R., & Miller, C.A., 2004. Trust and Etiquette in High-Criticality Automated Systems. *Communications of the ACM* 47(4): 51–55.

Parasuraman, R., Sheridan, T.B., & Wickens, C.D., 2000. A Model for Types and Levels of Human Interaction with Automation. *IEEE Transactions on Systems, Man, and Cybernetics Part A: Systems and Humans* 30(3): 286–297.

Porcelli, A.J., & Delgado, M.R., 2009. Acute Stress Modulates Risk Taking in Financial Decision Making. *Psychological Science* 20(3): 278–283.

Royakkers, L., & van Est, R., 2010. The Cubicle Warrior: The Marionette of Digitalized Warfare. *Ethics and Information Technology* 12(3): 289–296.

Salas, E., Cooke, N.J., & Rosen, M.A., 2008. On Teams, Teamwork, and Team Performance: Discoveries and Developments. *Human Factors* 50(3): 540–547.

Sharkey, N., 2010. Saying "No!" to Lethal Autonomous Targeting. *Journal of Military Ethics* 9(4): 369–383. Available online at http://www.tandfonline.com/doi/abs/10.1080/15027570.2010.537903 (accessed 28 January 2016).

Sheridan, T.B., & Verplank, W.L., 1978. *Human and Computer Control of Undersea Teleoperators* (Report ADA057655). Arlington, VA: Office of Naval Research.

Sparrow, R., 2007. Killer Robots. *Journal of Applied Philosophy* 24(1): 62–77.

Tversky, A., & Koehler, D.J., 1994. Support Theory: A Nonextensional Representation of Subjective Probability. *Psychological Review* 101(4): 547–567.

Woods, D.D., Tittle, J., Feil, M., & Roesler, A., 2004. Envisioning Human–Robot Coordination in Future Operations. *IEEE Transactions on Systems, Man and Cybernetics – Part C: Applications and Reviews* 34(2): 210–218.

Yaniv, I., & Kleinberger, E., 2000. Advice Taking in Decision Making: Egocentric Discounting and Reputation Formation. *Organizational Behavior and Human Decision Processes* 83(2): 260–281. Available online at http://www.ncbi.nlm.nih.gov/pubmed/11056071 (accessed 28 January 2016).

8 Civilizing Drones by Design

Aimee van Wynsberghe and Michael Nagenborg

1. Introduction

Unmanned aerial vehicles (UAVs), or drones, have come to dominate military applications in the 2000s. These remotely operated aircraft allow for the benefit of targeted killing without the same risks to combatants given that combatants do not have to be physically present for the killing to take place. These drones, however, have many more capabilities outside of killing at a distance that are also exploited for the benefit of non-military applications. In particular, they are thought to provide use for work that is "dull, dirty or dangerous" (Finn and Wright 2012). According to the European Commission, there are now more than 400 projects in 20 European countries seeking to develop drone technology for use outside the military (Reuters 2013). Accordingly, in recent years we have seen an increasing concern about the use of military technologies, drones among others, in civilian contexts voiced in academic writings (e.g. Graham 2011), as well as in public debates[1] (e.g. the militarization of police forces in the United States has been a hot media topic at the time of writing, August 2014). These concerns about the militarization of police forces represent an aspect of the larger challenge presented by so-called dual-use technologies: a blurring of the intended values from one context to another. In this chapter, we will look at drones as an example of a dual-use technology and explore the relationship between their design, their use, and the resulting values expressed in a new (and different) context of use.

We acknowledge that the framing of drones as military technology might be misleading in parts. Yet we will take this common perception as an entry point. More precisely, we will explore specific ethical issues related to the design of drones as their use shifts from a predominantly military context to the civilian domain. In this chapter, we will argue that theories such as value-sensitive design (VSD) are capable of providing us with a better understanding of why dual use raises ethical issues.

In the first part of our chapter (Section 2), we will present a short overview about the origins and reasons of the ethical concerns about dual use. We will then present the embedded values and VSD approaches as a way of understanding the relationship that drone technology shares with the realization of values. With this foundation, we can then explore how drones fit the value hierarchy of a military context and how drones may or may not fit the (differing) value hierarchy of a civilian context.

In Section 3, we will present an overview of likely capabilities of drones used by the military. The goal at this point is to identify the values embedded in the (drone) technology by the very capabilities of the drone. To do this we use the approach of van Wynsberghe (2013)[2] that seeks to link the consequences of using a technology with the expressed value. This method follows the overall approach of value-sensitive design but seeks to retrospectively align design choices regarding capabilities with their impact on the promotion of values when the technology is used in context. For this paper, we will use the capabilities of drones in a military context as a way of sketching the values embedded into the drones. We suggest, given the desire to use this technology, that these resulting values are those intended by the engineers or, at the very least, are desirable for the context of use. If they were not, we would presume that the technology would not be as successful.

Following this, we will explore how the likely capabilities of military drones meet, or fail to meet, the functional requirements of different civil sectors. In this section, we will discuss how drones may be (re)designed in order to increase their acceptability for these non-military contexts. We will explicitly *not* argue that any of these changes make drones more desirable. Rather, applying insights from VSD theory may contribute to making drones less problematic once their contexts of use are shifted.

Finally, we will have a look at the interplay of design and regulations as two interrelated strategies to make the use of drones more acceptable in the civil context. Again, we are not making any claims about the desirability or even the feasibility of using drones in these contexts. We simply argue that *if* there are good reasons for making use of drones in a specific context, there are *at least* certain demands on the design as well as on the regulatory frameworks within which they are employed.

This chapter has multiple aims and objectives. To begin, many scholars believe that engineers and designers bear a certain amount of responsibility for the outcome of their labor, and it is with this insight that engineers are encouraged to consider the unintended consequences of their work. The so-called responsibility gap (see the Introduction in this volume) seems to present a challenge to this general assumption. But on the one hand, even the acknowledgement of the challenge might not stop us from expecting engineers and designers from acting in a responsible way (Nagenborg et al. 2008). On the other hand, many aspects of the designs of drones are not related to this specific challenge of adaptive systems. In the following, we will mostly focus on these features. For these reason, we wish to contribute to the discussion of drones used in civilian contexts in order to ensure that due diligence occurs. We hope to engage in a conversation of the complexities in ethically addressing dual-use technologies as well as the conversation related to the particular ethical issues associated with the use of drones outside the military. In so doing, we aim to provide a method for (or at the very least a starting point for a method) exploring dual-use technologies using VSD.

Given the current state of affairs in which drones are available to the public at large in addition to the police sector, we believe this chapter is quite timely. One of the many benefits in exploring the concept of dual-use with respect to drone technology lies in contributing to a discussion of responsibility as it applies to the designers, engineers, and policy makers of drone technology. We suggest that part of this responsibility requires exploring and making explicit the unintended uses, contexts, and users. In certain instances, the results from such a study must be fed back to the designers in the hopes of building in a design consideration to mitigate the foreseen negative consequences (van den Hoven et al. 2012).

2. Dual Use and Value-Sensitive Design

2.1 Military-civil Dual Use

To begin, it is important to point out three things: First, the phrase "dual use" can mean either used by a dual class of people, i.e. military vs. civilian use, or it can refer to the same tool being used for good purposes or bad purposes. We refer to both dimensions in this work. Our main aim is to use the concept of dual use to discuss drones as conceived by, and used originally in, the military domain but shifting from this context of use to the civil domain. We also use it to explore whether this shift in context makes it a bad or good use. This is done through a value analysis (Miller and Selgelid 2008).

Second, it has been proven to be extremely difficult to make a clear-cut distinction between military technologies and non-military technologies. Third, we need to remind ourselves that dual use means dual income for those who develop and sell these technologies. Limiting the options for dual use is therefore not in the "natural" interest of the developers and producers of these technologies. We must also acknowledge that "military drones continue to earn most of the money, [and therefore] most drones in development still resemble military hardware" (Rothstein 2015: 41).

Dual use, in reference to technologies, has been around since the 1970s and traditionally referred to civilian technologies shifting to the military context. Of late, dual use has come to refer to a distinction more broadly construed: technologies that were designed and developed for one user, use, or context that are later used by a different user, for a different use, or in a different context. Consider a robotic exoskeleton intended and designed to assist combatants carrying heavy loads over long distances. The same (kinds of) robot(s) are now being used in rehabilitation centers to aid patients with spinal cord injuries in walking.

What makes this line of thinking problematic is the underlying assumption that there is a clear-cut criterion that allows us to differentiate non-military technologies from military technologies. The Internet is a good example to point to the complexity of the problem: one may point to the Arpanet as one of the historical origins of the Internet. Since the Arpanet was funded by the U.S. Department of Defense, it could be classified as military technology. Yet how do we take into account the shaping of the Internet by academics, hackers, hobbyists, industry, and others?

Perhaps it is a question about the source of funding to label something as a military technology; that is, if the military pays for the research and development of a product, it may be referred to as a military technology. However, as we have seen in the preceding examples, once the technology becomes pervasive outside a military context, the source of funding for initial research and development may no longer apply, given the continuous sources for research and development. What's more, technologies may cross the border between the defense sector and civil sector more than once. We may refer to this phenomenon as bidirectional dual use.

Finally, it should not be overlooked that everyday technologies may play an important role for the military. Take, for example, the Toyota Hilux, a pickup truck that has been called "vehicular equivalent of the AK-47" (Somaiya 2010) because of its common use in asymmetric warfare (Ammicht Quinn and Nagenborg 2013). Should we therefore refer to vehicles like the Toyota Hilux as military technologies?

From the standpoint of ethics, there is a close link between the responsibility of scientists and engineers and the debate on dual-use technologies. After the development and employment of the atomic bomb, it became questionable whether the works of scientists and engineers could be understood as being merely instrumental and value neutral. The atrocities committed in Nazi Germany and other totalitarian states in the name of 'science' added to the need of rethinking the idea of science as a neutral means of knowledge production and technology as a mere tool. In the aftermath of World War II, interest in the Hippocratic Oath was renewed and extended beyond the scope of medicine to other disciplines. One of the earliest examples is the suggestion by Gene Weltfish made in an article in *Scientific Monthly* (1945): "I pledge that I will use my knowledge for the good of humanity and against the destructive forces of the world and the ruthless intent of men; and that I will work together with my fellow scientists of whatever nation, creed or color, for these, our common ends" (Weltfish 1945, quoted in Ammicht Quinn and Nagenborg 2013).

The current debate on dual-use technologies often implies that it is easy to make a distinction between military technology and non-military technology. As we have demonstrated, it's very hard to draw a line between these two kinds of technologies. In the following, we would like to demonstrate that VSD theory can help us to avoid the issue of how to differentiate between these two kinds of technologies by looking at particular artifacts in specific contexts. Furthermore, it helps us make clear the relationship that dual use shares with ethics.

2.2 Embedded Values and Value-sensitive Design

Since the early 2000s the fields of human–computer interaction and ethics of technology have concerned themselves with the relationship between values and design. From one perspective, there are values embedded into a technology that are brought into existence through the use of the technology. This is referred to as the embedded values theory (Nissenbaum 2004; Brey 2012; van Wynsberghe 2013). Take for example privacy-enhancing technologies (PETs) like Privacy

ABCs (Attribute-based Credentials) or the company Silent Circle. In both cases, the situation is such that when using the product, the value of privacy is a consequence of using the technology; in other words, the value of privacy is embedded into the technology. For Privacy ABCs, using cryptography technology, individuals' identities can remain anonymous when interacting with others, thus promoting privacy. When using Silent Circle, individuals can have untraced and untracked phone calls, thus promoting their anonymity and privacy.

From the embedded values standpoint, a proactive theory and method to design has been developed named value-sensitive design (VSD) (Friedman 1996; Borning and Muler 2012; Friedman et al. 2013). VSD is an approach to design that systematically accounts for human values and stakeholder considerations throughout the design process. There are conceptual, technical, and empirical phases of the VSD approach, each aimed to explore a divergent dimension of the relationship between values and design. The conceptual phase is dedicated to uncovering and understanding the values at stake from an ethical/philosophical standpoint, as well as in the context of use. The empirical phase is aimed at any number of stakeholder issues (i.e. direct and indirect users) related to the values: generating a list of values of ethical import, critiquing a list of values of ethical import, prioritizing values, identifying value dams and flows for users, and so on. The technical phase is dedicated to understanding the artifact in context and how it manifests values or fails to do so. The phases are iterative, and the methods for engaging in each of the phases are multiple, varied, and developing from one moment to the next. It is an exciting time for the researchers of VSD as the literature builds and promising results gather. Although criticisms of VSD must be overcome, what is clear for researchers in this area is that making values explicit and working towards their manifestation is a promising avenue to designing good products for society and to establishing trust between designers and users.

The consequences of this theory are multiple and significant for engineers, designers, and policy makers: first, there is a normative dimension to the work of designers in that the values of a society are promoted or limited by a specific design. Second, with this in mind, the designer holds a certain amount of responsibility to estimate the consequences and embed a strategy to mitigate anticipated negative consequences. Returning to the Silent Circle example, one ought to consider that, although the technology is intended to be used by vulnerable or marginalized individuals or groups in dangerous situations who wish to protect their identity, the same technology could also be used by terrorist or other malicious users for terrible uses (i.e. the technology has the possibility for dual use with potentially drastic consequences).

The question for designers may be, "How can we design to ensure that only the intended users have access?" The question for ethicists might be, "How can, and should, we evaluate this dual-use dimension of the technology?" And, the question for policy makers might be, "How can we limit the use of the technology to the intended users?" All three kinds of questions deal with the tension between the intended values of designers and those that are (or may be) realized in a context of use.

Of course, all this being said, one may still wonder why the idea of embedded values raises concerns about military–civil dual use? The simple answer to this is: context matters. A context like the military is governed by values dissimilar to those in a domestic or policing context. Added to this, a context like the military may have values comparable to those of a domestic setting, but they may be interpreted or prioritized differently. One might consider, for example, that the military is governed by strict rules and laws that maintain order in an otherwise unstable setting – militants are not allowed to point their weapons directly at a person unless they are forced to by the activity of that person and intend to shoot that person. What's more, warfare demands a kind of objectivity and distance, whereas policing demands that officers do their job not at a distance but as members of the community.

2.3 Exploring Values in Design

With this in mind, let us now turn to an exploration of the capabilities of drones used in military contexts and the values that these capabilities reflect. The manner in which we do this reflects the approach of van Wynsberghe (2013) in her work on care robots. Although our focus for this chapter is on drone technology, we believe the approach of van Wynsberghe is still applicable for a variety of reasons. In particular, the approach targets the relationship between embedded values and robotics technology. More specifically, the approach fosters a method for unpacking values in design, exploring roles and responsibilities of humans and nonhuman artifacts, and the relationship that practices in one context share with the overall aims of the domain (e.g. what drones have to do with the overall goals of the military). Additional reasons will be discussed later in the chapter.

The methods of this approach involve a series of consecutive steps: data collection, analysis, comparison, and recommendations. For a retrospective evaluation, one must collect information regarding a specific practice, context of the practice, actors involved, and what values are manifest through the actions and interactions of actors. The practice is then analyzed according to the manifestation of values: what, how, and why values that come into being are ranked and interpreted in the ways they are. Once this is complete, the robot technology is then described to give a picture of the technical features (e.g. appearance, interface), capabilities (e.g. locomotion, infrared sensing, etc.), and level of robot autonomy (e.g. robot requires human input to execute its task; robot is fully autonomous for travel but requires input at a certain stage; etc.). Once this is complete, the practice is once again analyzed, only this time with the inclusion of the robot. The variety of practices (i.e. with and without the robot) are then compared in terms of the value analysis to make recommendations. Depending on the stage of development and the target audience, the recommendations will differ in their complexity and impact.

The method for this chapter will reflect the approach of van Wynsberghe but will also differ in many respects. First, in the same way in which van Wynsberghe seeks to uncover the values embedded in robot technology through a retrospective evaluation of the robots, we will list the capabilities of drones and uncover the values they help to bring into existence. This relationship is dictated by the fact that the consequences of using the technology determine the values in question. Thus, we look to the consequences of using drones in a military context in order to derive the intended values. To be clear, this may not be the epistemic aim of the engineers, but we search for what values are brought into existence and assume a minimal correlation between this and the aims of designers.

Second, just as van Wynsberghe argues for a design-by-design analysis – that is the same robot used in different contexts must have a different ethical evaluation in light of these differing contexts – we too will undertake a kind of design-by-design analysis. This is evident in how we attempt to assess the same class of robots, unmanned aerial vehicles, for multiple contexts. Our aim is to investigate how the same technology is capable of promoting or threatening values depending on its context of use. This will be investigated through multiple contexts of use.

Where this approach differs from the work of van Wynsberghe is that we are not comparing different kinds of care robots; instead, we focus on unmanned aerial vehicles or drones. Added to this, we are not using a specific ethical theory for an evaluation of drones. We are not attempting to make a claim as to how design choices were justified or could be justified. Rather, we are using the theory guiding VSD as a means for uncovering and making explicit the complexity of values related to the dual-use aspect of drone technologies (i.e. drones used outside the military). Lastly, we are not addressing individual practices and their complexity; rather, we are focusing on distinct contexts and the value landscape of said contexts.

This work is also meant to contribute to the growing field of VSD, in particular the practical applications of VSD. We aim to show that VSD as a theory can be applied for more than just future design but also as a tool for retrospectively evaluating current technologies based on the relationship between values and design choices.

3. Drones in Context

We begin our analysis of drones in context by exploring the capabilities of drones differentiated between one context and another. In the following sections, we will provide data on the capabilities of drones, the manner in which they are used (i.e. the practices for which they are used), and the relationship this technology shares with the larger picture (i.e. the military domain). What we present here is our basic understanding of the use of drones in the defense sector.

We will discuss the distinction between the military and policing (or domestic) contexts with respect to the use of drone technology as a way of highlighting issues of responsibility, governance, and regulation of drone technology.

3.1 The Military Context

In this section we will focus on the likely capabilities of drones used in a military context. We are aware that such a list of capabilities may be considered misleading because it homogenizes UAVs used in armed conflicts. Further research might be needed to go beyond this overview; however, for the purpose of our chapter, we consider the following overview to be sufficient as it provides us with a list of likely features of military drones that may not be seen as acceptable for drones in a civil context.

To begin with, we need to ask why UAVs are considered to be desirable for military operations. In his book *War X: Human Extensions in Battlespace* (2005), Tim Blackmore offered the following simple yet effective model. When discussing the "story of tank design", he points to the:

> triangle of mobility (speed), weight (armour), and firepower (tank cannon plus ammunition) . . . . Somewhere in the triangle is a balance point between the three. To make a tank fast means surrendering weight in armor and firepower; a well-protected, heavily armored tank makes for sluggish movement, one that can [be] wiped out by more agile weapons. . . . Properly drawing the triangle of speed, armor, and firepower will result in what the U.S. Army calls 'survivability' – both the weapon and its operator will survive encounters with the enemy.
>
> (Blackmore 2005: 72)

Removing the human operators from within the war machine allows for a different design: when survivability is reduced to the survivability of the machine, there is no longer the same need to shield the human bodies inside. The triangle can then be redrawn. This now has significant consequences for military drones: they no longer need to protect a human inside.

It is equally important to consider that removing the human operator from the inside of the "metal box," be it a tank or an aircraft, means having overcome the limiting factors of the human body – functional bodily needs. Drones are capable of travel over long distances and staying airborne for a long period of time. This allows a different design of war machines and new ways of fighting wars.

Removing the human operators from the inside also allows drones to become micro and mini. The shrinking of war machinery fits well with the tendency of drones to remain invisible to the enemy: "UAVs are designed to stay in the dark and bring the enemy into the light . . .; they are formed for seeing and not being seen; . . . are fitted to have knowledge and deny it to others" (Blackmore 2005: 137).

The ways in which UAVs allow their operators to 'see' and 'know' relates to three more characteristic capabilities of military drones. First, any reference to 'seeing' actually may be misleading because UAVs don't have eyes; they have sensors. Cameras gathering data from the portion of the electromagnetic spectrum that is visible to the human eye are just one kind of sensor. Because of these sensors, drones might be "flying above the clouds while tracking down our digital footprints" (Morais 2014). Consequently, there seems to be no reasonable expectation of privacy on the battlefield: "We really do want an Orwellian future, not in Manhattan, but in Kabul" (Huber and Mills 2002, quoted by Graham 2006: 257).

Second, UAVs must be understood as parts of a larger networked system. The data gathered by UAVs may not only be projected on the monitors of its human operators but may as well – and is likely to – be used in creating "real-time 'situational awareness'" (Graham 2006: 251). This further implies that the use of UAVs requires a communication infrastructure that allows the drone to connect to the network and to its human operators.[3] As Jutta Weber (2009: 90) has pointed out, these communication infrastructures are likely subjects of interception, since there is a trade-off between the security of the communication channels and the speed of communication when it comes to high-volume data transfers.

Third, it must be taken into account that the use of drones allows for seeing from a distance. In other words, the user of the UAV is at a distance from the battleground. Once the human body has been removed from the inside the machine, the distance between the UAV and its human operator seems to have become less relevant: satellite links allow for remote control of UAVs over many thousands of kilometers (Weber 2009: 96).

Relating all these capabilities (i.e. the consequences of having such capabilities) to the language of values, we may observe many pragmatic and ethical values at stake such as but not limited to survivability, safety, security, privacy, efficiency, efficacy, and competence. Removing the human operator from inside the war machine has allowed for a different design approach to this military technology. In particular, the value of survivability still remains an important value to be realized, but now the survivability of the vehicle itself (with the notable exception of disposable UAVs) is foremost. Added to this, it has become possible to overcome some of the limitations of the human body (by removing the human body), allowing the machinery to stay airborne for a longer period of time and to shrink in size (becoming micro and/or mini in some instances). The values of efficiency, safety (of the human operator), and endurance are realized through this. This, in turn, has made drones less visible to the enemy, while at the same time rendering the enemy visible by use of sophisticated sensor technology in combination with network capabilities; that is the values of efficacy and competence are implied here. Finally, drone technology allows for creating a distance between the operator and the targeted area. The function of teleoperation allows the realization of additional security (and safety) for the operator yet also changes the relation between the combatants, of which some are now removed from the battle space.

3.2 Shifting from Military to Civil Domains

In this ection, we continue our data collection on drones used in contexts, only this time we will look at the (current and future) use of drones in different civil domains: agriculture, rescue and disaster management, policing, and city governance. This is not meant to be an extensive list as we will not be able to take into account the whole variety of drones that may be used in these specific contexts. Rather we will ask what kinds of capabilities of military drones (as laid down in the previous section) make it desirable that drones are used in these contexts and how some of these capabilities will question the acceptability within that specific context. At the same time that we analyze the scenarios in which drones may be and are being used, we engage in a preliminary comparison between the military context and the civil context in question. In the concluding section, we hope to make recommendations for policy development (i.e. implementation and regulation) and for future studies.

3.2.1 Agriculture

In agriculture, UAVs are used to survey crops, dust crops, and spray pesticides over an area. The capabilities of drones mirror those listed for drones used in military contexts, including removing the human body, facilitating a greater distance between user and the work field (in this case the farm rather than the battleground), and sensors for acquiring environmental cues. In short, the drone is intended to fulfill the dull and/or dirty tasks associated with crop management.

In his article on "Agricultural Drones", Anderson (2014) sums up some of the advantages of using drones in terms of sensing capabilities:

> Drones can provide farmers with three types of detailed views. First, seeing a crop from the air can reveal patterns that expose everything from irrigation problems to soil variation and even pest and fungal infestations that aren't apparent at eye level. Second, airborne cameras can take multispectral images, capturing data from the infrared as well as the visual spectrum, which can be combined to create a view of the crop that highlights differences between healthy and distressed plants in a way that can't be seen with the naked eye. Finally, a drone can survey a crop every week, every day, or even every hour.
>
> (Anderson 2014)

For many, the use of drones in agriculture is not (at least at first glance) problematic. When we consider the values expressed through the use of drones in this context, we may uncover values such as efficiency, competence, and security. We suggest efficiency, given the speed at which the drone can work and the amount of information collected during that time, because the system may be considered a more efficient means for conducting these tasks. We suggest competence, given that the drone is capable of exposing information that is otherwise not available at eye level; thus, the drone fulfills tasks with greater competence, thereby increasing the overall competence of the farmer. With this in mind, we suggest that the value of security is manifest. Once in possession of this heightened level of information, the farmer is able to make decisions with a greater quality and quantity of information to secure the well-being of the crops.

What also makes the use of drones in agriculture look acceptable is that fields are well-defined, somewhat isolated geographical areas. One might suggest that the geographical location is able to limit privacy concerns from the beginning. Still, drones may need to take off from a distant position and therefore might be able to collect data that is not related to the fields or its task. Here, advanced privacy-sensitive sensors could be useful – and may, as we will later argue, be considered a (future) legal requirement. For example, sensors may become active only over the field where the drone is supposed to be operational and may acquire data only from that field. Redesigning sensor devices in accordance with privacy-by-design principles has become a common topic in the discussion on so-called Smart CCTV systems. Hence, it seems reasonable to make use of a similar approach for the design of drones with advanced surveillance capabilities. In this context, there is no need for capabilities to collect digital footprints or for the drone to be a part of a larger network. We would suggest, then, that a drone in this context requires only the necessary capabilities for the task at hand. In other words, drones for agriculture need only the capabilities to survey crops (rather than individuals), to crop-dust, and to do this with the human removed from the machine and able to work from a distance.

On further analysis, one may question the role and responsibility delegated to the drone in these instances, given that the technology is now accountable for a large portion of the responsibilities traditionally associated with the role of the farmer. It is also possible to suggest that farmers may lose certain skills associated with their roles. For these concerns, we suggest that as long as the farmer remains in the loop and is essentially responsible for all decisions related to the actions of the drone (i.e. the farmer dictates when crop dusting happens, when images are taken, etc.), then concerns of responsibility may be appeased. With this suggestion comes the presupposition that even if the technology were capable of making such decisions, they should remain within the domain of the farmer.

3.2.2 Rescue and Disaster Management

In this type of scenario, we imagine the use of drones for dangerous situations in which the human needs to be protected from the environment. Current examples of this kind of drone include, for example, the OpenRelief project, which strives to "support the emergence of open, humanitarian drone technology" (OpenRelief 2015), as well as the use of the Global Hawk system to map the terrain after the earthquake in Haiti in 2010 (Freeman 2014).

Drones in this instance must have sensors to allow them to collect geographical information of an area and also to detect the presence (or absence) of humans in the area. The drone is required to work without a human on board (to protect the pilot from the hazardous environment) and is also required to enable the human to work from a geographical distance (in the same way as when the drone is used for agricultural purposes). The drone is expected to work in unpredictable and unstructured environments. To do this, the drone must be a part of a larger network to communicate and coordinate with the users on the ground.

Given these capabilities, we suggest that the values of safety, security, and competence are strongly implied. When keeping human operators off the field, we consider the value of safety to reflect the protection of civilians from the dangerous environment. We consider the value of security to refer to a stabilization of the dangerous and/or unpredictable environment in case humans must go in afterward. Or it is possible that the drone engages in surveillance tasks to allow distance operators to make decisions about additional security measures. We again refer to the value of competence in that the drone may be capable of collecting information beyond the capabilities of humans (i.e. the drone may have infrared sensing capabilities for detecting bodies). We believe competence to be of not only pragmatic but also of ethical significance, especially in this application, given the delicacy of the context (e.g. one wrong move may end in lives lost) and the severity of the potential consequences (e.g. lives lost, injuries, and/or further damaged incurred).

One basic challenge should be mentioned (as with all rescue and disaster management technologies): at first glance, there seems to be no need to prioritize privacy. It is very tempting to argue that nobody will give priority to privacy if/when in need of help. However, the systems must be fully operational before the disaster has happened; hence, so-called mission creep is a big issue here – and very tempting, especially when it comes to policing. Thus, for these applications, drones need to be visible to prevent limitations on privacy and, above all, because low visibility is not a value in disaster relief actions.

3.2.3 Policing

Another prominent potential domain for the employment of drones is policing. "Police forces use (unmanned aircraft systems) to monitor large crowds, prevent or detect crime and assist in incident responses" (Finn and Wright 2012: 188).

Since policing includes such a large variance in tasks, including dull, dirty, and dangerous tasks, there is some overlap with the application of drones in this area and the other domains previously discussed. Policing can also be understood as an important part of city governance, which has many branches and tasks. In this section, we will distinguish between dangerous policing tasks and those that are dull or mundane.

For these two applications, the drone capabilities involve removing the human body from engaging in dangerous tasks, facilitating a greater distance between user and the work field (in this case a protest, riot, dangerous neighborhood, crime scene, and so on), and sensors for acquiring environmental cues from the work field. When replacing the police officer(s) for dangerous tasks, values such as safety, security, competence, reliability, impartiality, accountability, accuracy, and provenance are involved. Safety here refers to the safety of the policy officers.

Given the time at which this chapter was written (2014–2015) and the scandals of police brutality in the United States against African American males, it may also be of interest to explore the idea that the use of drones to replace police officers may also be in the best interests of citizens. For these reasons, we also list the values of reliability, impartiality, accountability, accuracy, and provenance. These values are related to one another in that the drone may provide a realistic portrayal of incidents that does not rely on witness or victim testimony, thereby promoting values such as impartiality (not favoring one side over another in the dissemination of data), reliability (the data can be depended upon), and provenance (the collection and ownership of data can be accounted for).

When replacing police officers for dull tasks (as opposed to dangerous tasks) such as patrolling neighborhoods, responding to less urgent calls, and so on. In such instances, the drone may be considered a tool for surveillance, a tool that excels those capabilities of the already existing CCTV cameras for security purposes, for example. Of importance here is that drones differ from the existing technology in that they are capable of collecting data from multiple angles (i.e. they are not stationary) and may not always be observable. Consequently, they promote the value of competence, but this may be done at the expense of citizen privacy.

On closer inspection, it is necessary to uncover value tensions that may be relevant for the ethical permissibility of the use of drones in this context. For this, we must look at the capabilities of increased mobility and teleoperation as well as the network capabilities. As worries of militarization are voiced with regard to the use of military technology by police, we need to be aware of the specific differences between policing and war fighting. An obvious difference here is that members of the police forces (hopefully) cooperate with citizens, while war fighters take on the enemy.

This has important implications for the value of teleoperation, which potentially removes the police officer from the field of operation and limits opportunities to interact with citizens on the ground. Hence, one of the key challenges to be met when deploying drones for policing is to avoid weakening the ties between policing agencies and the communities they operate in. In such cases, one might suggest that drones may be used only for limited purposes (of course, the next difficult task is to delineate these limits).

Teleoperation might also be seen as problematic if the operator has only limited knowledge about the community where the drone is being employed. This in turns makes it harder for the operator to interpret the activities monitored. In short, there is a need to recognize the tension between the value of community policing and the other values previously listed that is manifest through the use of UAVs in policing (i.e. replacing officers for dangerous and/or dull tasks). This value of community as it relates to policing is the reason why many of us feel relief to see a police officer in a time of need or feel compelled to seek the assistance of a police officer for asking directions. In Western countries at least, police officers are not traditionally considered antagonistic to the community but are considered trusted members of the community, helping to protect citizens and to maintain a safe and secure environment. Many might argue that such ideals can be accomplished only when the police are considered members of the community, as opposed to outsiders there repress citizens.

While this tension may not be fully resolved by redesigning drones, increasing the visibility of the UAVs might at least help us to realize the value of accountability, which is seen as a key value in policing. Added to this, increasing visibility of a policing UAV may also help to balance the value of privacy in relation to the values of safety and security. At the very least, citizens should be able to recognize a policing drone as such. Hence, the capability of UAV limited visibility needs to be questioned when it comes to this specific civil context.

Following along the lines of privacy concerns, network capacities may also create worries about potential privacy issues and more importantly 'social sorting', or issues of discrimination.

> This overview demonstrates that while UAS devices have been used in a range of applications, it is the same "usual suspects" who are targeted by UAS surveillance. Eick argues that in Western Europe, there is "hardly a marginalized group that is not targeted by UAVs", and this paper illustrates that this is common to other countries as well. Large crowd monitoring generally focuses on protesters, "hooligans" and "anti-social" elements.'
>
> (See Finn and Wright 2012)

The case of crowd monitoring is difficult to describe as a dangerous policing task and is just as difficult to describe as a dull policing task. Depending on the country, city, event, and crowd, the task could be either dull or dangerous. One need only consider the difference between crowd monitoring at the peaceful protests in Hong Kong circa 2013 compared with the crowd monitoring tasks of the historical protests such as the Tottenham protests circa 2011 in the UK or the Soweto protests circa 1976 in South Africa. Very little can be done to predict (with great accuracy) the nature of a crowd gathering. The question of importance for designers and policy makers of drone technologies is how to balance the values at stake for crowd monitoring when one cannot predict whether this is going to be a dull or a dangerous task. In particular, one might suggest that the value of privacy may not be compromised when the event is dull, but in cases when the event is dangerous, the safety of citizens and officers take precedence over privacy concerns. At the same time, one must also consider that the limiting the value of privacy may have negative impacts on the safety and security of citizens. In both instances we would claim, as a starting point, that the values of accuracy, accountability, reliability, and provenance are of the utmost importance and should be prioritized accordingly.

What is important to remember here is that (a) drones are employed for surveillance of the usual suspects and (b) advance sensor technologies and network capabilities allow singling out members of a crowd. The term "crowd monitor" might even be misleading as the focus is not only on the crowd as such. As a design solution, we would suggest that when using drones for early detection of mass panic, there is no need to single out any individual.

4. Concluding Thoughts: Shifting Contexts Require Analysis of Values and Intentions

The aim of this chapter was not to solve all of the issues related to the design and implementation of drone technologies used in contexts outside the military. Rather, the intention was to broaden the discussion of the use of drones outside the military in terms of a discussion of the intended values of engineers. An additional objective of this chapter was to show the complexity of value analysis and the utility of VSD methods for engaging in a discussion of the relevant values and the potential value tensions that may be encountered. In the preceding sections, we engaged in a preliminary evaluation of the use of drones in civil contexts by comparing value analyses from drones in military contexts with drones in a variety of civil contexts. We based our general methods for this approach on previous work done by van Wynsberghe, although we made some significant changes in our scope of analysis and the types of recommendations that can result from such an analysis.

Before we begin the substance of our concluding section, we want to pause here to recognize and discuss a morally relevant distinction between epistemic uses of drones (i.e. as surveillance) and action uses (i.e. lethal force).[4] The distinction to be made here concerns the use of drones for our ability to collect information and consequently to have knowledge about a situation or person (surveillance) vs. our ability to use the technology strictly for achieving a physical end; is either of these uses more problematic than the other? For the former, consider a future world in which drone surveillance records are necessary for proving the innocence or guilt of an offender, much like DNA evidence is now the gold standard. Will the technology become so pervasive that our norms and understanding of evidence will change based on their functions and roles? For the latter, consider a world in which human soldiers are never again sent to battle, but instead wars are fought through unmanned aerial vehicles. Perhaps the lives of soldiers would be saved, but who would then be the target for these drones – civilians, children, and the unarmed? These are not problems that a VSD approach can necessarily capture; rather, this is a discussion not for risk assessment but for ethics of technology on a larger scale. We do not engage with this distinction any longer but felt the urgency to make it clear and to define another limit to our analysis.

The conclusion of this chapter is simply to state that it is necessary to examine the relationship between military drones and design values in order to expand the analysis of drones as they now enter different (unintended) contexts. Our concern here has not been and is a question not of whether military drones are good or not but rather of how they are designed and what this design says about the values at stake. By exploring the relationship between military drones and embedded values, we come to understand how drones in a military context do and should embed values such as safety, security, and competence. These values are directly related to the capabilities of drones, e.g. low visibility, "military at a distance," signal transmission with low levels of encryption, and minimal levels of privacy. These are the general values needed to make the technology applicable for a military context. The question of interest for this paper, then, has to do with whether these values match up when the technology is used in an alternate context.

As we have seen, the technology of drones when used in a civil context, for instance agriculture, maintains many of the same capabilities (with the exception of advanced network capabilities) but promotes the values of efficiency, competence, and security. In fact, we may go so far as to suggest that the drone enhances the abilities of farmers to fulfill their roles according to these values. We were also explicit to point out that this application does not come without its concerns; there may be potential negative outcomes if one were to consider a deskilling of farmers or a loss of connection between farmers and their traditional practices.

The application of drones in policing raises issues related to officers working at a distance, the visibility of individuals, and the privacy of citizens in general. The practice of policing is characterized for many as a strengthening of the community – that police officers are part of the community. Of particular interest is the question of privacy: in the military context, information may not require the same degree of privacy given that it is shared among militants; in a policing context, information about citizens may be subject to higher degrees of privacy.

Given these insights into the value tensions of UAVs as the technology moves from one context of use to another, we suggest a way to mitigate these concerns through regulations of the technology. Drones with certain capabilities (that embedded certain values) can and should only be used in a context of use that conforms to that value hierarchy. In other words, in order to justify the use of a drone in a given context, its capabilities and the corresponding embedded values must conform to the value hierarchy of the given context.

This brings us to the question concerning the relation between design and regulations. As pointed out at the beginning of the chapter, dual use means dual markets. In order to maximize profit, it is quite tempting to sell the same kind of UAVs for different contexts of application and avoid the additional costs of any redesign that goes beyond the mere minimum. After all, if the same drone can be used in disaster relief and policing, then why should there be two variations of the same UAV on the market? Specific regulations on the use of drones in different contexts may be required to create the kind of incentive needed to civilize drones by design.

Of course, the idea of asking for specific regulations is more likely to be accepted in a European context rather than in the U.S. context. Yet we need to keep in mind that there are already legal regulations with regard to robots (Nagenborg et al. 2008), as well as specific regulations for specific domains where UAVs might be used. These regulations might need to be reconsidered once we have good reasons to propose the use of UAVs in these contexts. Hence, a debate on the regulation of drones is something that can hardly be avoided in the first place. What we are arguing for here is to consider domain-specific rules that take into account the different design properties of UAVs. Rather than trying to address privacy concerns by regulations only, we suggest to limit the surveillance capabilities of drones to the minimum needed for the drone to function in a specific context. As Adam Rothstein (2015) has pointed out, "[E]ven the legality of a drone's data collection cannot guarantee that this aspect of drones will always function as designed" (Rothstein 2015: 94). Yet this cannot be a justification against the application of *privacy-by-design* principles to limit capabilities. It seems rather reasonable to ask for domain-specific regulations that include references to set limitations on the capabilities of the UAV, as well as the realization of domain-specific values such as visibility and accountability.

The offshoot of our discussion is that the acceptable use of specific drones in specific contexts requires a strong legal framework that might not be in place in all states. This at least suggests that the export of these technologies should be strictly regulated, if not outright banned. It may also indicate that the development of these kinds of drones should be stopped.

Notes

1 As an example, the HBO television series "Last Week Tonight with John Oliver" has a dedicated episode on the militarization of police forces in the United States in addition to the exploration of police brutality in Ferguson, Missouri, United States of America.
2 We also draw on the methods suggested in van Wynsberghe and Robbins (2014) and Nagenborg (2013).
3 Tjerk de Greef (Chapter 7 in this volume) offers an insightful discussion of the challenges presented by teleoperated drones and networked decision-making.
4 As de Greef (Chapter 7 in this volume) already highlighted this: even within the military domain, UAVs are "considered a strategic asset for surveillance and reconnaissance tasks." For a detailed discussion of drone surveillance and how it relates to other forms of surveillance, refer to Wall and Monahan (2011) and Bracken-Roche et al. (2014), as well as the already previously mentioned works of Graham (2006) and Finn and Wright (2012).

References

Ammicht Quinn, R., and Nagenborg, M. 2013. Was ist "zivile" Sicherheit?, in R. Ammicht Quinn (ed). *Sicherheitsethik*. Springer, 253–263.

Anderson, C. 2014. Agricultural Drones. *MIT Technology Review*, Special: 10 New Breakthrough Technologies 2014. Available online at http://www.technologyreview.com/featuredstory/526491/agricultural-drones/.

Blackmore, T. 2005. *War X: Human Extensions in Battlespace*. University of Toronto Press.

Borning, A., and Muller, M. 2012. *Next Steps for Value Sensitive Design*. Proceedings of the ACM SIGCHI Conference on Human Factors in Computing Systems, 1125–1134.

Bracken-Roche, C., Lyon, D., Mansour, M.J., Molnar, A., Saulnier, A., and Thompson, S. 2014. *Surveillance Drones: Privacy Implications of the Spread of Unmanned Aerial Vehicles (UAVs)*. Surveillance Studies Centre, Queen's University.

Brey, P.E. 2012. Anticipatory Ethics for Emerging Technologies. *NanoEthics* 6(1): 1–13. doi:10.1007/s11569-012-0141-7.

Finn, R.L., and Wright, D. 2012. Unmanned Aircraft Systems: Surveillance, Ethics and Privacy in Civil Applications. *Computer Law & Security Review* 28: 184–194.

Freeman, W. 2014. *How Drones Are Revolutionizing Disaster Relief*. Thinkprogress (posted on 27 May 2014). Available online at http://thinkprogress.org/climate/2014/05/27/3440753/drones-disaster-relief/.

Friedman, B. 1996. Value-sensitive Design. *Interactions* 3(6):16–23.

Friedman, B., Kahn, P.H., Borning, A., and Huldtgren, A. 2013. Value Sensitive Design and Information Systems. Early Engagement and New Technologies: Opening up the Laboratory, in N. Doorn, D. Schuurbiers, I. van de Poel, and M.E. Gorman (eds), *Philosophy of Engineering and Technology*. Springer, vol. 16, 55–59.

Graham, S. 2006. Surveillance, Urbanization, and the U.S. 'Revolution in Military Affairs', in D. Lyon (ed), *Theorizing Surveillance: The Panopticon and Beyond*. Willan, 247–269.

———. 2011. *Cities under Siege: The New Military Urbanism*. Verso.

Huber, P.W., and Mills, M.P. 2002. How Technology Will Defeat Terrorism. *City Journal*, 12(1): 24–33. Available online at http://www.city-journal.org/html/12_1_how_tech.html.

Lemos Morais, R. 2014. Sky High, Skin Deep. Dark Technologies of Mediation, in A. Kroker and M. Kroker (eds), *Theory beyond the Codes*,15 September 2014. Available online at http://www.ctheory.net/articles.aspx?id=731.

Miller, S., and Selgelid, M. 2008. *Ethical and Philosophical Consideration of the Dual-Use Dilemma in the Biological Sciences*. Springer.

Nagenborg, M. 2013. Ethik als Partnerin in der Technikgestaltung, in Regina Ammicht Quinn (ed), *Sicherheitsethik*. Springer, 241–252.

Nagenborg, M., Capurro, R., Weber, W., and Pingel, C. 2008. Ethical Regulations on Robotics in Europe. *AI & Society* 22: 349–366.

Nissenbaum, H. 2004. Privacy as Contextual Integrity. *Washington Law Review* 79(1): 119–158.

OpenRelief. 2015. *About Us*. Available online at http://www.opendawn.com/openrelief/about/.

Reuters. 2013. *Airshow – Sky's the Limit for Civil Drones*. Reuters, 19 June. Available online at http://www.reuters.com/article/2013/06/19/air-show-drones-civilian-idUSL5N0EM1NJ20130619 (accessed 28 January 2016).

Rothstein, A. 2015. *Drone*. Bloomsbury Academic.

Somaiya, R. 2010. Why Rebel Groups Love the Toyota Hilux. *Newsweek*, 14 October. http://www.newsweek.com/why-rebel-groups-love-toyota-hilux-74195 (accessed 28 January 2016).

van den Hoven, J., Helbing, D., Pedreschi, D., Domingo-Ferrer, J., Gianotti, F., and Christen, M. 2012. FuturICT – The Road towards Ethical ICT. *The European Physical Journal Special Topics* 214(1): 153–181.

van Wynsberghe, A. 2013. Designing Robots for Care: Care Centered Value-sensitive Design. *Science and Engineering Ethics* 19(2): 407–433.

———. 2014. A Method for Integrating Ethics into the Design Process of Robots. *Industrial Robot* 40(5).

van Wynsberghe, A., and Robbins, S. 2014. Ethicist as Designer: A Pragmatic Approach to Ethics in the Lab. *Science and Engineering Ethics* 20(4): 947–961.

Wall, T., and Monahan, T. 2011. Surveillance and Violence from Afar: The Politics of Drones and Liminal Security-scapes. *Theoretical Criminology* 15(3): 239–254.

Weber, J. 2009. Robotic Warfare, Human Rights & the Rhetorics of Ethical Machines, in R. Capurro, M. Nagenborg, and G. Tamburrini (eds), *Ethics and Robotics*, IOS Press, 83–103.

Weltfish, G. 1945. Science and Prejudice. *Scientific Monthly* 61(3): 210–212.

Part IV

Drones and Moral Responsibility

9 Drones, Automated Targeting, and Moral Responsibility

Alex Leveringhaus

1. Introduction

The issue of automation is an important emerging topic in the debate on drones.[1] By 'automation', I mean that a machine (not just a drone), once programmed, is capable of carrying out predefined tasks without assistance from an operator. Many drones are already automated to some degree, for example to enhance flight performance. Arguably, targeting is the most controversial area of automation, and I therefore focus on it here. More precisely, I investigate to what extent, if any, automated targeting affects the assignment of moral responsibility to drone pilots and drone programmers, respectively. By 'assigning moral responsibility', I mean that one credits or debits an agent for an event. Some commentators argue that automated targeting makes the assignment of moral responsibility difficult because it reduces drone pilots and other so-called cubicle warriors to mere 'marionettes' (Royakkers & van Est 2010). While this is an exaggeration, I will show that automated targeting in drones poses subtle challenges for moral responsibility and that some of these are best met, I contend, by what I call a 'Standard of Care Approach'.

2. Background Considerations

In this part of the chapter, I outline three sets of background considerations that are crucial for my subsequent analysis.

2.1 Automated Targeting

Automated targeting within drones is not an either/or issue. Put differently, it would be a fallacy to conceive of drones as either having automated targeting functions or not having automated targeting functions. When thinking about targeting functions, the model of a five-stage decision-making loop, or a Kill Chain, is particularly useful (see Leveringhaus & de Greef 2014).

1 *Observation stage*: The acquisition of information about a particular target or a specific situational scenario.
2 *Orientation/analysis stage*: Analysis of the available information.
3 *Decision stage*: Making a targeting decisions based upon the analysis of the available information at stage 2.

4 *Enactment stage*: Enforcement of a targeting decision made at stage 3.
5 *Assessment stage*: Assessment of the aftermath of the military act.

Automation can occur at *any* stage of the Kill Chain. For example, drones have complex sensor suites and can record vast amounts of data. The resulting information could be too much for a single pilot to process. In response, stages 1 and 2 may be automated, so the pilot sees only images that have already been selected by the drone. This could potentially enhance the pilot's performance during stage 3. Once a targeting decision has been made by the pilot at stage 3, the delivery of the payload stage 4 may be automated again. Clearly, this does not exhaust the possible options for automation in the targeting process. But it illustrates that automated targeting is a complex issue.

To cope with this complexity, my analysis operates with two major distinctions. Firstly, I distinguish between drones with *partially* automated targeting processes and drones with *fully* automated targeting functions. In the former, only certain steps of the Kill Chain are automated, while, in the latter, *all* steps of the Kill Chain have been automated. Secondly, I distinguish between drone pilots and drone programmers. The former teleoperate drones with partially automated targeting functions and, crucially, are actively involved in the third and fourth steps of the Kill Chain. I refer to those who programme drones with fully automated targeting functions as drone programmers rather than pilots. Of course, pilots and programmers might interact with each other. This is because programmers might be needed in order to automate certain steps within partially automated Kill Chains. I return to this issue shortly.

2.2 Moral Responsibility

In their introduction to the present volume, Santoni de Sio and Di Nucci helpfully distinguish among three types of responsibility, which, though interrelated, must not be conflated: *causal*, *legal*, and *moral* responsibility. In this chapter, I am interested only in moral responsibility, though this is not to say that automation does not raise any legal issues.[2] My underlying account of moral responsibility is roughly Strawsonian in character (Strawson 1962). This means that it steers clear of the intricate discussions of the topic found in the literature on free will. Firstly, for the Strawsonian approach, moral responsibility must be understood in terms of social practices, especially practices of praising and blaming. Secondly, when an agent is morally responsible for an event, the agent is liable to praise or blame for that event, and other agents are justified in forming appropriate reactive attitudes towards him or her. The general question this chapter pursues is whether the increasing automation of drones necessitates a rethinking of practices of praising and blaming.

2.3 Just War Theory

Any practice-orientated account of moral responsibility requires a moral framework against which practices of praising and blaming operate. In this chapter, I approach the relationship between moral responsibility and automated targeting from the perspective of just war theory. Just war theory, in its current incarnation, operates with two regulatory frameworks that restrict the use of armed force, as well as one framework that addresses the aftermath of war: (1) *jus ad bellum* (justice in the declaration of war), (2) *jus in bello* (justice in the conduct of war), and (3) *jus post bellum* (justice after war). Assignments of responsibility play an important role in all three frameworks. In this chapter, I am interested in whether automated targeting makes it more or less difficult to assign responsibility to drone pilots and drone programmers for transgressions of *jus in bello*'s three main criteria:

1 *Distinction*: The distinction criterion obliges belligerents[3] to distinguish between legitimate and illegitimate targets by not intentionally targeting the latter.[4]
2 *Proportionality of means*: The proportionality of means criterion obliges belligerents not to cause excessive harm.
3 *Military necessity*: The military necessity criterion obliges belligerents not to cause unnecessary harm.

Before I press on, let me raise two general points about *jus in bello*. Firstly, note that assignments of moral responsibility for violations of *jus in bello* principles are also relevant to *jus post bellum*. In order to be able to deal with the aftermath of war, one needs to be able to assign responsibility for military acts *in* war. While I shall not explore this point in this chapter, it is worthwhile bearing in mind that the implications of automated targeting extend beyond the actual conduct of war and into its aftermath. Secondly, drones have attracted attention because of their use for so-called targeted killings during counterterrorism operations (see Finkelstein, Ohlin, & Altman 2012). My view is that *jus in bello*'s three principles are just as relevant to counterterrorism operations as they are to 'regular' or 'conventional' armed conflicts Thus, in what follows, I shall not differentiate between automated targeting during peacetime, international armed conflicts, or non-international armed conflicts. For the law, these distinction are crucial, but, arguably, they are less relevant to just war theory.

3. Moral Perception, Moral Choice, and Automated Targeting

3.1 Responsibility after Nuremberg

As pointed out earlier, moral responsibility has something to do with social practices of praising and blaming. However, while such a practice-oriented account is a helpful starting point, it is not sufficient to make sense of responsibility in the military. Roughly, to be considered a *responsible* agent and thus liable to praise and blame, an agent needs to fulfil the following three conditions (Cowley 2014: 26–28). Firstly, the agent needs to have the *moral capacity* to understand what he did and why he is being held responsible for his actions. Secondly, the agent needs to be able to *morally understand* the situation in which he acted. Thirdly, the agent needs to have *control* over his actions; more precisely, he must have the ability to do otherwise. It is, of course, one thing to say that morally responsible agents fulfil these conditions. It is quite another to show what it means to fulfil them in different contexts. Arguably, the pressures an agent faces in the course of armed conflict are different from those experienced during peacetime.

To make sense of moral responsibility in war, the Nuremberg trials serve as a useful historical guide. Prior to Nuremberg, soldiers, in order to be exculpated from wrongdoing, merely had to show that they had followed a properly authorised order. Nuremberg changed this by introducing two additional criteria (May 2005: 181–182). Firstly, the *moral perception* criterion demands that soldiers show that they had attained or could not have attained morally relevant knowledge that allowed them to assess their use of armed force in a particular situation. Secondly, the *moral choice* criterion demands that soldiers show that they could have avoided the execution of a particular order. While the moral choice criterion roughly reflects the aforementioned requirement that an agent must have control over his actions, the moral perception criterion is more complicated.

Firstly, the moral perception criterion presupposes that a soldier is *morally competent*. That is to say, the soldier must know the relevant moral framework that underpins her search for facts. Reflecting the *jus in bello*'s distinction criterion, the moral perception criterion rests on a rule that the intentional killing of innocent individuals, however defined in detail, is impermissible: this is something a morally competent soldier understands (May 2005: 182). Secondly, a morally competent soldier would then have to acquire all the necessary facts that allow him to assess a situation. Note that this step is necessary but not sufficient to have adequate moral perception of a situation. Thirdly and directly related to the preceding point, a soldier needs to identify those facts that are *morally* relevant in terms of the moral rules that she knows. For instance, the fact that a person neither poses an unjust threat nor can be held responsible the creation of such a threat is morally relevant because it renders that person innocent and thus morally immune to intentional attack. If a soldier knows all the facts arising from a specific situation and can identify those that are morally relevant, the soldier is not only morally competent but also morally *aware*. In short, to fulfil the moral perception criterion, soldiers needs to be morally competent *and* aware.[5]

Viewed through the lens provided by the moral perception and moral choice criteria, drone pilots and drone programmers could argue that they should be exculpated from wrongdoing because automation has either undermined their moral perception or deprived them of control over their actions. The metaphor of a marionette underlines this worry. Marionettes, at the very least, lack control over what they do. But is it really sound to view drone operators and drone programmers as marionettes? If the answer is yes, this would constitute a powerful argument against certain types of drone technology. For automation, it is to be feared, would essentially take us back to a pre-Nuremberg world. The pre-Nuremberg defence that 'I am not to blame for wrongdoing because I only followed orders' would be replaced by a similar defence: 'I am not to blame for wrongdoing because I only followed what automated mechanisms told me' or, 'I am not to blame because the drone did it'. The stakes, then, are high.

3.2 Automation and Moral Competence

Let me begin with a general point about moral perception. As we just saw, moral perception has two components, moral competence and moral awareness. One question is whether automation undermines the moral competence of those who operate drones with partially or fully automated targeting functions. This charge is particularly pertinent in the case of drone pilots who teleoperate drones, reflecting the claim that teleoperation engenders a so-called video-game mentality, or at least moral disengagement (Sharkey 2014). But it could also be levelled at drone programmers who never actually see or otherwise encounter the enemy. If war resembles a video game or if the enemy is just an abstract 'other', distinctions between different categories of individuals may appear less relevant to soldiers. This could undermine the understanding that drone pilots and programmers have of the basic *jus in bello* rules that govern the use of armed force.

As I said earlier, critics who level this charge at drone technology have not convincingly backed up their claims. I do not think that teleoperation on its own threatens the moral competence of drone pilots. Nor does the programming of fully automated drones on its own threaten the competence of drone programmers. The methods of remote drone warfare might, in fact, have the opposite effect in the sense that it removes one stress factor that might make soldiers 'snap': living in close proximity to the enemy and feeling threatened by the enemy. The case of Staff Sergeant Robert Bales, who, after four combat tours of duty, 'snapped' and murdered 16 Afghan civilians, may serve as a chilling reminder of a soldier who lost his moral competence – and for whom, as a result, *any* Afghan was fair game (BBC News 2013).

In general, moral competence implies that, as we just saw, an agent understands basic moral rules, such as the prohibition to target the innocent. Automated targeting does not necessarily challenge developing adequate moral competence. Whether a member of the military develops or fails to develop adequate moral competence depends, I contend, much more on training than on subsequent uses of a particular weapon. Consider the following case. In a racist state, soldiers are taught that black skin colour is a legitimate targeting criterion and that anyone who happens to be black is a legitimate target. Skin colour, of course, is not a legitimate criterion for targeting. Now suppose that the racist state's drone pilots practise teleoperated targeting exclusively on black people or that drone programmers exclusively practise the automation of targeting procedures by using GPS coordinates of, say, black ghettos. All this is used to subtly reinforce that any black person is a legitimate target. Something has clearly gone very wrong here. But this has nothing to do with drones or with automated targeting. The problem would remain exactly the same if the racist state trained, say, its police officers who go out on the beat in the very same manner: black people are fair game.

Sound legal and ethics training, however rudimentary it might be for some members of the military, is key to sound moral competence. Morally competent drone pilots and programmers would be trained to understand the basic principles of *jus in bello*. Like any other soldier, they should understand which criteria are relevant to targeting and which are not. Automated targeting neither hinders nor enhances this learning process. We can assume, then, that a morally competent drone pilot would know that, as in our example, black people are not legitimate targets because they are black. A drone programmer would know that he has to abide by the principle of distinction when programming a drone with targeting coordinates. Programming a drone in order to deliberately attack civilian areas is forbidden. Of course, drone pilots and programmers may know and understand the relevant rules. The challenge is to apply them. In order to abide by the rules of combat, drone pilots and programmers need to know the relevant facts for targeting. Thus, when considering the impact of automated targeting on the moral responsibility of drone pilots and drone programmers, let us see whether it undermines their moral awareness rather than their moral competence and whether it restricts the choices they can make as a result.[6]

3.3 Facts and Choice

Let us look at the relationship between moral awareness and the partial automation of targeting first. As I argued earlier, automation can potentially affect all five stages of a teleoperated Kill Chain: (1) observation stage, (2) orientation/ analysis stage, (3) decision stage, (4) enactment stage, and (5) assessment stage. From the perspective of the moral perception criterion, automation within the first two stages of the Kill Chain are particularly problematic. Drone pilots can have the necessary moral awareness only if they acquire the morally relevant facts that enable them to make a morally sound decision that complies with *jus in bello*. The automation of stage 1, however, can edit out some information that might have been crucial for the assessment of the situation in stage 2. As has been seen, automation of stage 1 is potentially a response to the large quantities of information that teleoperated drones can record via their sensor suites, and that would be impossible for a human operator to process. Yet the editing out of certain information, while enabling drone pilots to deal with the sheer quantity of it, may not necessarily enable human operators to develop adequate moral awareness. Note, though, that this is not necessarily an argument against automating stage 1. This is because not automating stage 1 does not always lead to better moral awareness: too much information might equally hinder the acquisition of *morally* relevant facts.

While automation at the early stages of the Kill Chain might hinder the acquisition of morally relevant facts, automation of the latter stages of the Kill Chain can impact negatively on the moral choices that drone pilots are capable of making. Automation of stage 3, for instance, restricts the kinds of targeting decisions that pilots can make. An automated systems may, for instance, allow them to make only certain targeting decisions. This is not necessarily a bad thing, though. Restricting the choices available to drone pilots can also restrict the scope for human error or the abuse of the system. Automation of stage 4 might make it difficult for drone pilots to stop the application of force to a particular target. Again, it does not follow that automation of stage 4 is impermissible. An automated process may deliver a payload with greater accuracy and security than a human operator could.

The picture that emerges is complex. There is no doubt that automation can have a negative impact on the moral perception of drone pilots, as well as the control they have over their actions. That said, automation also has its redeeming features. It is just not guaranteed that, once automation is taken out of the picture, it becomes easier for drone pilots to fulfil the moral choice and perception criteria.

While automation may have redeeming features in partially automated Kill Chains, fully automated Kill Chains are more morally problematic. This is because, compared to drone pilots, drone programmers have little control over what the drone does, once it has been deployed. Of course, it is possible to argue that drone programmers should be given the power to override the system. By designing an override mechanism, drone programmers retain the capacity to stop the drone and can thereby prevent transgressions of the moral rules of war. In this sense, just as a drone pilot retains the capacity not to press the button on the joystick, drone programmers could be said to fulfil the moral choice criterion: they can override the original programming of the drone.

It is questionable, however, whether it is realistic to expect that drone programmers will be able to prevent misapplications of force by overriding the drone. They might, in particular, lack the morally relevant facts to assess whether an automated drone is about to wrongfully engage a target. Programmers are neither in the area where the drone operates, nor do they, unlike drone pilots, necessarily control the drone via teleoperation. If this is correct, then we have jumped from the frying pan into the fire. While an override function would give drone programmers some control over the drone, they lack the adequate moral awareness that would allow them to exercise effective control. As a result, they are unlikely to fulfil the moral perception criterion. This shows that moral choice and moral perception are closely related. Without adequate moral awareness, soldiers may retain the capacity to do otherwise, but they will not be inclined to exercise that capacity.

In general, the problem for the moral choice and perception criteria is that future, fully automated drone warfare – and robotic warfare more broadly – might be conducted at a speed and over distances that make it difficult for human individuals to override the drone, once it has been deployed. Given the speed involved, the cognitive abilities of human operators may not be sufficient to attain relevant moral awareness of a situation. It is possible to contend that this is a powerful reason against the further development of drones with fully automated targeting functions. I am sympathetic to this view. But rather than following this argument, I am content with showing that this does not necessarily undermine the moral responsibility of drone programmers. Rather, by devising sound standards of care, we can ensure a commitment to responsibility within the armed services.

4. Standards of Care for Automated Targeting

As has been shown, the partial and full automation of targeting processes has advantages but also pitfalls. One way of dealing with the latter consists, I think, in broadening our understanding of moral perception. At the moment, moral perception is defined in rather narrow terms. It requires soldiers to ensure that they have morally relevant knowledge that enables them to distinguish between innocent and non-innocent individuals in war or between legitimate and illegitimate targets. At first sight, this does not sound like much. The (apparent) lack of demandingness reflects the stressful situations in which soldiers operate. However, with the rise of drone technology, some of those stresses have been removed. Those who operate drones, either as pilots or programmers, are not exposed to the same pressures as an infantry soldier. There is thus no reason why, in their case, the demands of moral perception should not be broadened.

More precisely, moral perception should not merely be confined to the perception of a situation *during* deployment. As we just saw, especially in the case of drone programmers, this is not necessarily possible. Instead, adequate moral perception should be centred on the perception of risks that may arise from the use of automated targeting functions during the deployment of a drone. As I have observed on a different occasion, the notion of risk is currently underexplored in just war theory. Yet it is crucial for the treatment of drones, as well as for other types of automated weaponry. The use of automation can certainly have benefits. But this should not distract us from also appreciating the risks posed by automation, not least the risks that result from the limited potential for overriding an automated drone once it has been deployed. I cannot outline a procedure that enables us to assess in risks in warfare. Suffice it to say here that riskless warfare is not possible. So different risks need to be balanced against one another.

If this is sound, drone pilots and drone programmers are responsible for the imposition of risks on others by relying on automated targeting procedures. Surely, they can be held responsible for imposing the resulting risks, even if their capacity for moral awareness and control is restricted during the deployment stage. It is precisely this restriction that leads to risk. A morally competent and aware drone pilot or programmer would recognise this and take steps to minimise risks accordingly. Moral perception is crucial for how we think about risk in warfare. Whether drone pilots and programmers, following the deployment of an automated drone, were morally aware of risks and gave adequate consideration to the risks can make a difference as to whether the imposition of risks is *justified*, *negligent*, or *reckless*. If the risk is justified, its imposition would not be morally wrong. Conversely, the negligent or reckless imposition of risk signals that a drone programmer was not morally aware or perhaps even lacked moral competence. If a drone pilot or programmer did not take adequate steps to attain the morally relevant facts that enabled her to assess the risks arising from automation, she would be guilty of a serious wrongdoing. David Rodin suggests that negligence in war may morally be on a par with terrorism, though I would not necessarily go so far.

The argument that risks must be adequately perceived by drone pilots and programmers reveals a further facet of moral responsibility. Many assignments of responsibility are backward-looking. That is, they assign responsibility to an agent for an event that occurred in the past. Recall the earlier example of an automated drone whose deployment caused a disproportionate number of civilian casualties. In this case, the assignment of responsibility to a drone programmer would automatically involve a past event. However, responsibility can also be *forward-looking*, something that is not always captured by the practice-orientated approach. In other words, there can be responsibilities to prevent a future event. In this sense, the extension of the moral perception criterion is forward-looking because it seeks to prevent future transgressions of the rules of war that may result from excessive risk taking.[7]

The broadened moral perception criterion gives rise to the wider idea of a Standard of Care Approach. This approach stresses the importance of defining sound standards of care for the use of partially or fully automated targeting functions in drones. It would delimit specific contexts in which automated drones could be deployed, imposing a reasonable, as opposed to an excessive, risk on others. It is futile, in my view, to condemn the use of automation in drones as such. Whether automation should be used, as well as to what extent it should be used, depends on particular contexts and their related standard of care. Assignments of responsibility to drone pilots and drone programmers take place against the background of standards of care. We can justifiably blame drone pilots and programmers for working with an inadequate standard of care or for not abiding by an existing and morally sound standards of care. Note that the resulting standard of care would also govern interactions between drone pilots and drone programmers. To reduce risk, drone programmers would have to be transparent about the ways in which they programme partially automated drones. They would have to inform their colleagues about the parameters being used for automation, as well as the stages of the targeting process being automated.

It is, at this stage of the argument, important to stress that the idea of a standard of care does not apply solely to those individuals who directly operate drones but also to their superiors. It is the responsibility of military superiors and of the institution of the military more generally to devise appropriate standards of care for new and emerging technologies, such as automated drones, as well as to review existing standards of care and update them.

In sum, automated targeting processes within drones challenge social practices of blaming in the military. They deprive drone pilots and drone programmers of some aspects of their agency. But not all is lost. I argued that, in order to be able to assign responsibility to drone pilots and programmers, existing criteria, such as the moral perception criterion, need to be revised. It is important to realise that practices of praising and blaming, as well as the moral frameworks against which such practices operate, are not static but subject to change. Automation may challenge *existing* frameworks, but this does not mean that these frameworks cannot be updated. Two issues, I argued, are important. Firstly, adequate perception of risks resulting from automation is absolutely crucial, and this should be highlighted by militaries seeking to develop a sound standard of care for the use of automated targeting processes. Secondly, the notion of risk serves as a bridge between two understandings of responsibility, a retrospective understanding, which is close to social practices of blaming (for past events) and a prospective understanding, which is not fully captured by practices of blaming. Essentially, the prospective, or forward-looking, understanding of responsibility leads to a wider standard of care that seeks to prevent future wrongdoing.

5. Praising 'Warrior Geeks'?

In the preceding parts of the chapter, I analysed to what extent automation impacts on practices of blaming military personnel for transgressions of the laws of war. However, practices of 'praising' occupy the other side. It is often forgotten that the attribution of responsibility to an agent involves not only blaming the agent for what she has (has not) done but also, depending on the circumstances, praising the agent for an act or omission. Societies have developed practices for praising agents for desirable behaviour: heroism, kindness, charity, and so on. There is probably no other institution in society where social practices of praising are as prominent and important as in the military. Soldiers are routinely praised for displaying martial virtues of physical courage, sacrifice, stoicism, loyalty, patriotism, and professionalism. One question, which I tackle in this final part of the paper, is whether practices of praising retain their relevance in an age of increasingly automated drone systems.

Those critical of the military could point out that practices of praising have long been out of step with technological and social developments. For instance, given that, in an age of post-heroic warfare, high-altitude bombing is often employed, not least by Western militaries, some might question whether it is still appropriate to praise fighter jet pilots for their acts. It is possible to defend practices of praising in this context by pointing out that fighter jet pilots need to display some courage. Flying a fast jet in itself requires courage, regardless of the likelihood of being shot down by the enemy. Furthermore, in order to carry out their tasks, fighter jet pilots need to endure tremendous physical and mental challenges during training. Though fighter jet pilots do not necessarily exhibit the same courage as, say, infantry soldiers about to advance on enemy positions, it might not be entirely inappropriate to praise them for their conduct.

While it may be (just about) possible to uphold practices of praising in the context of fighter jet pilots, the case of drone pilots seems to pose intractable problems. Neither physical courage nor physically arduous training is required to teleoperate a drone. Nevertheless, I do not think that it is inappropriate to praise drone pilots. The hole left by the absence of physical courage can be plugged by mental courage and resilience. True, teleoperation does not entail physical risks, but it can entail mental ones. Unlike fighter jet pilots, drone pilots, even if some processes underpinning teleoperation are automated, are directly confronted with the aftermath of an attack. There are some indications that drone pilots develop symptoms of post-traumatic stress.[8] Furthermore, precisely because of the absence of a physical threat, we can expect adherence to a higher standard of compliance with the laws of war – and more restrictive standards of care – from drone pilots than from the infantry soldier. In light of this, we can praise drone pilots for their professionalism and their willingness to incur mental risks.

Arguably, this line of argumentation becomes difficult to sustain in the case of drone programmers. Like drone pilots, drone programmers do not need to display physical courage or endure physical hardship during training. Unlike drone pilots, however, drone programmers are shielded from the bloody aftermath of an attack. The case of programmers may highlight what Christopher Coker (2013) calls the rise of the "warrior geek": the whizz kid capable of programming a complex machine while being insulated from any of the havoc he wreaks. Needless to say, if the Standard of Care Approach outlined in the preceding part of this chapter is sound, we must expect drone programmers to comply with an appropriate standard of care. We can certainly blame them for non-compliance with it. But it is less certain whether compliance on its own is sufficient to be the subject of praise. It might not be. As already observed, compliance with the law is often expected as a bare minimum of acceptable behaviour in civilian as well as military life. While drone programmers may exhibit great professionalism, this may be necessary but not sufficient to make them subjects of praise.

Yet it is possible to construct a consequentialist case in favour of praising drone programmers. Consequentialists typically claim that practices of praising (and blaming) are justified insofar as they have beneficial consequences; however, these may be defined in detail. Put simply, these practices keep people on the straight and narrow, encouraging desirable behaviour while deterring undesirable behaviour. Programming a drone is a difficult task, and the military has an interest in programmers carrying out the task to the best of their ability. If practices of praising ensure that drone programmers comply with a standard of care, the consequences of praising them are surely beneficial and thus desirable from a consequentialist perspective.

A potential counterargument, also from a consequentialist perspective, is that such an instrumentalist understanding of praising can potentially have negative consequences. For instance, those who incur physical and/or mental risks might feel alienated if drone programmers are praised for simply complying with standards of care. Potential divisions within the service between fighter jet pilots and drone pilots, on the one hand, and drone programmers, on the other hand, might open up as a consequence, undermining the efficiency of the service.

If these observations are sound, militaries that are investing in fully automated drones are caught between a rock and a hard place. On the one hand, without the prospect of being praised, drone programmers will not necessarily be motivated to abide by a high standard of care. On the other hand, praising drone programmers for instrumental reasons might undermine the cohesion of the service, which potentially can have negative effects on the willingness of military personnel to comply with standards of care.

If anything, this shows that the biggest challenge for highly technologically advanced militaries in the 21st century consists in integrating into the service those who teleoperate and, more importantly, programme uninhabited weapons systems, such as drones. Interestingly, then, and contrary to claims that uninhabited weapons technologies reduce their operators to 'marionettes' with diminished responsibility for military acts, the problem with automated (and teleoperated) drones is not that they make the attribution of blame difficult. The assignment of blame can potentially be handled via a sound standard of care. Rather, praising becomes problematic. Given the significant role played by praise in the military and in society's relationship with the military, this is no trivial matter. Whether existing practices of praising remain meaningful and relevant in an age of automated warfare remains to be seen.

6. Conclusion

It is reasonable to argue that drones are here to stay. Across the world, we are witnessing increasing investment into drone technology. Armed conflict in the 21st century is likely to involve the deployment of drones. Technology never stands still, and we can already see that automation is going to play a prominent role in the drones that are going to be developed over the next couple decades. Clearly, there are precedents for automation in the military. Future developments in automation will build upon those precedents. The automation of targeting functions in drones, either partially or fully, raises a number of important issues. In this chapter, I concentrated on those relating to the moral responsibility of individual drone operators, as either pilots or programmers. I made three main claims.

Firstly, insofar as practices of blame are concerned, automation necessitates rethinking what it means to be a morally responsible agent in the military. This can be done by developing sound standards of care for the use of automated targeting functions. The main message is that, while automation might challenge practices of blaming as they *currently* stand, nothing prevents a rethinking or updating of criteria for blame.

Secondly, I argued that responsibility is not only concerned with wrongdoing in the past but can also entail obligations to prevent wrongdoing in the future. Again, the development of sound standards of care is crucial in this respect.

Thirdly, I contended that responsibility is not solely focused on blame and the prevention of wrongdoing but also has a positive dimension. Practices of praising are integral to the operation of the military, yet they are challenged by further advances in automation. This poses considerable challenges to the coherence of the service, which militaries in the 21st century will have to confront.

In sum, the chapter has shown that, while drones may not be radically new and may be placed on a continuum with other precision-guided weapons, their introduction and further development are not unproblematic. Militaries need to adapt their practices of blaming and praising accordingly. On the one hand, drones do not reinvent the wheel of weapons technology. On the other hand, they indicate that things may not be able to remain as they currently are. In this sense, they signal both continuity and change.

Notes

1 For further treatments of issues arising from automation, see the respective chapters by de Greef (Chapter 7) and Koch (Chapter 4) in this volume.

2 Automation may also have repercussions for causal responsibility. Very often it is difficult to ascertain what exactly occurred. Imagine an intense exchange of gunfire between infantry soldiers. It might be impossible to ascertain exactly whose bullet killed an enemy soldier. Automation can potentially make a difference here. A blackbox within an automated drone, for instance, could record who programmed the machine,

as well as when and why the machine engaged a particular target. Assignments of causal responsibility are neither necessary nor sufficient in order to be morally responsible for an event. Nevertheless, more reliable assignments of causal responsibility may enhance investigations of wrongdoing in armed conflicts.

3 I use the term 'belligerents' in a broad sense. On the one hand, it refers to states and other organisations that are engaged in armed conflict. On the other hand, I also take it that it refers to those individuals who fight on behalf of states and non-state actors. *Jus in bello* imposes obligations on both individuals and collective/corporate entities.

4 There is considerable disagreement between just war theorists about who or what constitutes a legitimate target in war. I shall not delve into this debate here. See Walzer (2006); McMahan (2009); Fabre (2012).

5 I thank Cecile Fabre, Thomas Simpson, and Janina Dill for helping me clarify the concept of moral competence.

6 Critics could argue that the recent use of drones suggests a loss of moral competence. Certainly, the use of drones during counterterrorism operations is problematic because it targeting criteria have partly been redrawn, now including 'militants' (who are inadequately defined). I am not sure, however, whether this constitutes a loss of moral competence. The moral necessity to distinguish between those who may permissibly be targeted and those who are immune to attack is not questioned as such. Rather, counterterrorism operations illustrate that the appropriate application of *jus in bello*'s distinction criterion is highly contested.

7 For further reflections on risk, see Chapter 8 by van Wynsberghe and Nagenborg in this volume.

8 For some thoughts on PTSD in drone pilots, see Chapter 5 by Kirkpatrick in this volume.

References

BBC News. 2013. *Afghan Massacre Soldier Robert Bales Gets Life Sentence*. 23 August. Available online at http://www.bbc.co.uk/news/world-us-canada-23820790 (accessed 29 January 2016).

Coker, C. 2013. Warrior Geeks: How 21st Century Technology is Changing the Way We Fight and Think About War. London: Hurst & Company. Available online at http://www.amazon.co.uk/Warrior-Geeks-Century-Technology-Changing/dp/1849042543

Cowley, C. 2014. *Moral Responsibility*. Acumen.

Fabre, C. 2012. *Cosmopolitan War*. Oxford University Press.

Finkelstein, C., Ohlin, J.D., and Altman, A. (eds). 2012. *Targeted Killings: Law and Morality in an Asymmetrical World*. Oxford University Press.

Leveringhaus, A., and de Greef, T. 2014. Keeping the Human 'In-the-loop': A Qualified Defence of Autonomous Weapons, in M. Aaronson, W. Aslam, T. Dyson, and R. Rauxloh (eds), *Precision-strike Warfare: Strategic, Ethico-legal and Decisional Implications*. Routledge, 206–224.

May, L. 2005. *Crimes against Humanity: A Normative Approach*. Cambridge University Press.

McMahan, J. 2009. *Killing in War*. Oxford University Press.

Royakkers, L., and van Est, R. 2010. The Cubicle Warrior: The Marionette of Digitalized Warfare, *Ethics and Technology* 12: 289–296.

Sharkey, N. 2014. Saying 'No' to Lethal Autonomous Targeting, in T. Demy, G.R. Lucas, and B.J., Strawser (eds), *Military Ethics and Emerging Technology*. Routledge, 132–146.

Strawson, P. F. 1962. Freedom and Resentment. Proceedings of the British Academy, 48: 1–25.

Walzer, M. 2006. *Just and Unjust Wars: A Moral Argument with Historical Illustrations*. Basic Books.

10 Drones @ Combat

Enhanced Information Warfare and Three Moral Claims of Combat Drone Responsibility

Michael Funk, Bernhard Irrgang, and Silvio Leuteritz

1. Introduction

From a philosophical point of view, we are focusing on the moral responsibility related to concrete persons or groups of persons who are interacting with drones or talking about drones. But some issues of this chapter are also related to more general social and global challenges, which are calling for institutional solutions and can´t be addressed so easily to specific agents. We don´t discuss definitions of moral responsibility or ethical debates of the so-called responsibility gap, as this has already been done by Filippo Santoni de Sio and Ezio Di Nucci in the Introduction of this book. What we are going to do is a reflection of anthropological relations between moral responsibility and combat drones: linguistic, technical, and moral practice. Therefore, in a first step, our emphasis will be on the ways in which people talk about drones in terms of autonomy, intelligence, or responsibility. We differentiate three forms of language games and three forms of autonomy, and we discuss the so-called mirror-effect. Our conclusion is that, in order to think about moral responsibility, drones are most adequately described as "embedded technical autonomy", not in the meaning of human autonomy. In the second step, we discuss technological developmental paths, technical conditions for the usage of drones, and some characteristics of drones in relation to traditional combat technologies. We discuss aspects of the criteria of military success and introduce the term "enhanced information warfare" to describe both the dependence and advancement[1] of combat drones in relation to information technologies, microelectronics, and worldwide communication networks. Our conclusion at this point is that, in order to realize moral responsibility, drones must be treated as a global, transnational, and transcultural phenomenon, which calls for regulation by the UN or some other new institution. Thirdly, we are going to discuss fundamental aspects of moral responsibility, in search for an adequate ethos or moral codex of using, developing, or trading combat drones. We are arguing for a basic moral attitude in which idealistic aims must be combined with pragmatic skills. As a conclusion, we are going to summarize at least three fundamental moral claims (which could be complemented by others). Thereby, we are not going to argue for a ban of combat drones as such. But driven by the idea of human autonomy and enlightenment, we are searching for moral conditions for using drones responsibly.

2. "Embedded Technical Autonomy": Towards Linguistic Practice

Forms of warfare are changing dramatically. Classical warfare has been related to national states for centuries. Today's wars are different.[2] They become more and more asymmetric, shaped by terrorism and new technologies – such as micro-electronics, satellites, computer and information technologies, combat robots and drones.[3] In this chapter, our focus will be on drones as one of the latest forms of new high-tech warfare and their relations to moral responsibility from a philosophical point of view.

For Thomas Petermann and Reinhard Grünwald, technical consequences can be directly linked to machines on a causal level. But it is not possible to blame so-called autonomous technologies, machines, or automats – such as combat robots or drones – on a moral level. It is also not possible to punish tools for mistakes or moral misconduct. Also, the usage of "totally autonomous" technical systems is not justifiable as the law of *ius in bello* (law of war) stands (Petermann & Grünwald 2011: 198–202). We agree with that position. Therefore, in this first section, we are going to discuss Language Games of "autonomy". So let us ask, "What is the meaning of the word 'autonomy' in context of combat-drones?" Trying to answer this question in his paper on "Killer Robots", Robert Sparrow emphasizes a fundamental point when he states:

> In particular, it is unclear at what point a weapon system´s autonomy means that *it* is, in some real sense, making the decisions with which it is entrusted. An important reason for the lack of clarity about this question is that the nature and extent of autonomy, even in people, is itself a controversial and poorly understood matter.
>
> (Sparrow 2007: 65)

If we don´t know what "autonomy" means in human life, how can we use this word adequately to describe drones? Philosophical literature has been filled with controversial attempts to clarify meaning(s) of "autonomy" for centuries, and the debates are still going on. Problems of definition are one issue; another problem is related to the usage of words and cultural practices. On a basic linguistic level, we can learn from Ludwig Wittgenstein that meaning and notion are generated in the ways people use words in social interactions (Wittgenstein 2006). Here we generate Language Games and forms of life[4] as human practices. The semantic essence of words becomes secondary as meaning is interwoven into the ways people *use* words. We want to apply this insight to the practices of talking about autonomy in the context of combat drones.

Drones are a form of new and young technology, but the ways of using the word "autonomy" are older. We are confronted with a new technological tool that challenges us to develop adequate *technological*, *social*, and *linguistic* practice. If we develop new technological practice, we also develop new ways of talking and of socially interacting. In consequence, people are searching for ways to play new Language Games with already established words such as "autonomy". "Autonomy" remains the same word, but its treatment is going to be enhanced. Thereby we are used to linking the notion to specific implicit and tacit meanings of other words, such as “intelligence”, “creativity”, or “life” – but also “responsibility”. Nobody talks about autonomy isolated from those genuine contexts, even if those contexts shape the meaning implicitly and are not explicitly mentioned. So, if we play with the word “autonomy” related to a new technology, we tacitly try also to play with words like “intelligence”, “consciousness”, or “responsibility” in a new way. From a philosophical point of view, we can differentiate at least three ways in which those games are played. Usage of “autonomy” can be related to the following three contexts of intelligence:

> (1) The other human being as a person (with consciousness and awareness of his future, his death and himself): *Human Intelligence*;
> (2) Other biological intelligence like the intelligence of animals (with consciousness, possibly self-consciousness): *Natural Intelligence*;
> (3) Autonomous acting machines (without any model of themselves and without consciousness): *Artificial Intelligence.*
>
> (Irrgang 2014a: 40)

The form of human autonomy (1) is not the same as the form of drone autonomy (3) because the assumptions of intelligence in the ordinary linguistic context are not congruent, even when we use the same word – “autonomy”. If we talk about autonomous drones, we link the notion and usage of the word “autonomy” to artificial intelligence (form 3), not to human intelligence (form 1). The characteristic for form 1 is that people play those games – drones can´t play anything. Humans label combat drones as autonomous; drones won´t label themselves. If we have a new problem in defining the word “autonomy” today, it is not because of drones as such. It is because of the things people do with drones: practices in applying and talking about a new technology using old words. Thereby we often indicate those three forms without explicitly using words like “intelligence”, “autonomy”, or “responsibility”:

- “Tom will be able to handle the conflict on his own. Don´t worry, he’s a smart guy!” Language Games of “autonomy” can be related to human intelligence (form 1).
- “Look this cute and clever beaver! He is able to built the huge dam!” Language Games of “autonomy” can also be related to natural intelligence (form 2).
- “Have you seen the drone that is able to navigate inside a closed building without remote control? Wow, that´s high-tech!” Language Games of “autonomy” are also related to artificial intelligence (form 3).

But what happens if somebody asks, “Is the drone allowed to do that?” In this case, the meaning of human *moral* responsibility of human intelligence (form 1) is cross-linked to technological autonomy (form 3). The problem consists of a genuine “mirror-effect”: often people are mixing form 1 and form 3 unintentionally. They are talking about drones as if those technologies were humans or animals. Unconsciously drones are included in a human-related language practice.

This circumstance can be called a "mirror-effect" because by it we (humans) also want to understand who we (humankind) are. Drones are mirrors that confront us with implicit and tacit forms of linguistic, social, and technical practice. When we observe ourselves labeling drones as "autonomous", we can see a typical point of general robotics research: we try to figure out who we (humans) are while trying to recreate or rebuild ourselves.[5] Tools like drones enable us to think about our human fundamentals and anthropological characteristics. And on a cultural level, technical tools like drones belong to our environment and function as something like a material "mirror" of human practice.[6]

How we play with the word "autonomy" is a mirror effect itself: we are using an old word with different layers of meaning and apply it to a new entity (combat drones). Thereby drones are a mirror plate for the established practices of talking about autonomy. And at the same time our Language Games are a mirror for the practices in talking about and handling of drones. In other words, we are confronted with anthropological and social assumptions as we are forced to find words for what we do with those new technologies. That challenges not so much the logic or semantics of the word "autonomy" but mainly the many ways in which people are playing with it.

Also with respect to combat drones and responsibility, it is important to understand what "mirror" means in this specific context. If drones are mirrors, we should not reduce human autonomy (form 1) to technical autonomy (form 3). Metaphorically speaking, the drone as a mirror remains passive; it is a reflective plate – something like a "zeroth-person-perspective". But humans do something with a mirror: change bodily position, try gestures or make themselves up, and so on. Humans are multi-perspective beings, and that is the difference: human autonomy is a perspective phenomenon related to different forms of knowing and creativity (Funk 2014a; Irrgang 2005). It is not static: we do it. And that is also the point with combat drones: we do it – even if those new technologies change or enhance social structures or forms of warfare. Technologies are mirrors, but we are the guys standing in front of those mirrors. Only if people remain blind to themselves or are frightened of their own freedom, they may could find a non-perspective substitute for their own autonomy in drones or robots (form 3). But if people are self-confident, they must not be afraid because of their freedom, intelligence, creativity, and autonomy (form 1).

Focusing on the notion of responsibility, we can come to the same conclusion. Sparrow argues "that autonomy and moral responsibility go hand in hand" (Sparrow 2007: 65). That is true as people are used to playing Language Games of autonomy with an anthropological set of meanings that inherently also belong to the way in which we are used to talk about responsibility. When Sparrow relates autonomy to responsibility in the context of combat robots, he is playing a Language Game himself, in which he is confronting the form of human autonomy (1) with the form of technical autonomy (3). The confrontation of the two different categories leads to an aporia. Consequently, Sparrow argues:

> that the more these machines are held to be autonomous the less it seems that those who program or design them, or those who order them into action, should be held responsible for their actions. This itself is enough to suggest

> that the prospects for fully autonomous machines are more remote than is sometimes claimed.
>
> (Sparrow 2007: 66)

The point is that responsibility is related to the human form of autonomy (1). Anything else must be remote from a language-pragmatic point of view. It is hard to play a Language Game in which the term "responsibility" is linked to the technical form of autonomy (3), which is characteristic for drones. One conclusion could be that there are also different layers of responsibility. We could go on and postulate for every anthropological term another technical form that is also related to drones. But that would not make any sense. Doing so would devaluate our language. We generate meaning in practice, so we need adequate practice in talking about drones. We do not need any reductionist or simple technomorphism of ordinary language. We are humans; we are not drones. So we should talk anthropologically about humans but technologically about drones. "Responsibility" is associated with form (1). "Tools", "means to an end", or "technical function" is associated with form (3). Saying it with another metaphor: from a human point of view, drones are "semi-autonomous technologies" (Irrgang 2014b: 159f.).

Responsibility for the usage of drones is related to a concrete human person or groups of human persons such as operators, commanders, generals, or politicians – the superiors who control the tools or give orders to drone operators (see e.g. Funk 2014b and many others). With that definition, it is possible to understand drones as technical tools, which function semi-autonomously insofar they may fly or attack a target without any remote control or other human intervention. But the target itself, the aim, the goal, or the end of an operation – which involves the political, economic, or social background – is chosen and constituted by human(s). That is why persons or groups of persons are responsible[7] for drone operations, collateral damage, success, or failure. At the same time, the semi-autonomy of drones remains embedded into human means-end-oriented frameworks. Insofar, this form can be called more precisely "embedded technical autonomy". With that terminology, we play an adequate Language Game in describing drones.

Without a military or political order: no drone operation! Anything else would be criminal or self-destruction but not war. Embedded technical autonomy is not the replacement of political rationality or a substitute for political engagement at all. We could embed the usage of semi-autonomous drones in a framework of global peace – or at least in an honest try to reach global peace. Drones are new options for political decision-makers and become part of new global strategies in terms of security politics. Of course, like every tool, drones can also be misused[8] and embedded in unethical practice. But that, again, would mean that those tools are embedded in human (unethical) frameworks. So what we need to focus on is the human framework. Moreover, we want to argue that the basic ethical problem is not the creation of "totally human-like autonomous" drones. It is not possible to construct human-like drones, even if we use words like these. But the form of embedded technical autonomy, which seems to be realized at the moment and will be further developed in the coming years, will lead to massive ethical challenges regarding the frameworks. Therefore, it is not appropriate to argue against those technologies as such. In the following steps, we want to illustrate that the challenges of transnational and transcultural frameworks are more important. Drones will not replace humans or human responsibilities. But drones are new high-tech tools – mirrors that call for appropriate human solutions. Those solutions are related to traditional, cultural, and social backgrounds, while also demanding new steps at the same time. That is our hypothesis and will be emphasized in the next section.

So, if we talk about drones or other robotic systems, we should ensure that we talk about tools. Tools are always embedded in human means-end-oriented technical practice.[9] Technical practice is culturally embedded and a matter of implicit, embodied human knowledge.[10] That issue is relevantly related to drones, as those tools are included in global networks and raise questions of transcultural technical practice. This point will also be intensified in the next section.

3. "Enhanced Information Warfare": Towards Technical Practice

Processes of development are characteristic for technical tools. So called technological-economic developmental paths[11] constitute the genesis and use of technologies. There is no technology without any fore-form. Tools also have precursors. Besides social and economic factors, at least two historical issues shape the ways in which we are used to handling tools. The first is related to the material development of artifacts itself. For example, in information technologies many precursors of current personal computers can be found, such as the Commodore 64 and others. Today we may label those technologies as "outdated" or "far away from the technical state of the art". But from a historical point of view, Commodore 64 was an important step of technical development that enabled today's generations of computers to follow. Another example is related to scientific instruments such as telescopes. Galileo Galilei was using classical light-based telescopes in 16th- and 17th-century astronomical research. Our telescopes today are much more complex, some located in space, linked with supercomputers or catching x-rays. Current telescopes do not look like their 16th-century fore-forms. But historically a developmental path can be reconstructed, leading from Galilean telescopes to current high-tech instruments such as the Hubble Telescope.

The second impact of developmental paths is related to the familiarity and routines in which people use these tools. The focus is not on the materiality of the tool but on the skills of users. Many examples can be found in the arts: even if a piano and a clavichord remain constant on a material level, different pianists interpret scores in different ways. As soon as specific competences – techniques – are shared and taught to students, schools are constituted. Those schools frame developmental paths in handling pianos (Funk 2012; Funk & Coeckelbergh 2013). Another example is the geometry of computer keyboards. The genuine order of letters has been prefigured by mechanical typewriters because of mechanical reasons. After people had learned to use typewriters, attempts to introduce alternative geometries with computer keyboards were not successful. From a material-technical point of view, alternative orders of letters are easily realizable, but people were habituated to the typewriter structure and did not want to learn new skills in handling new keyboard interfaces. That is why computer keyboards and typewriters still have the same order of symbols: "Q-W-E-R-T-Y . . ." (David 1985).

Both aspects of developmental paths – traditions in constructing tools (material dimension of artifacts) and traditions in using tools (routine dimension of users) – can be found in military technologies as well. Without technical precursors, such as airplanes or rockets: no flying drones. Without space travel: no satellites. Without satellites: no GPS, no worldwide communication, no remote-controlled drones, and so on. The technological development of current weapon systems is also mainly a consequence of microelectronics and information technologies. These are key technologies, and a developmental path is leading from those technological structures directly to new war technologies such as combat drones. One consequence is a growing distance between operators and drones in combat situations. The bow and arrow, muskets in early modernity, the first airplanes and then cruise missiles in 20th-century warfare are examples of a developmental path of the growing distance[12] in military technologies.

In drone wars, face-to-face experiences are mediated by screens, which lead to specific epistemic and moral problems (Coeckelbergh 2013a, 2013b). Another consequence is related to information superiority. Somebody who wants to be able to operate a remote-controlled drone from a distance of thousands of kilometers evidently needs accurate information in order to succeed in the mission. That information superiority also enables potentials in differentiating between combatants and non-combatants on classical battlefields. Compared to classical cruise missiles, those new technologies could decrease collateral damage (Münkler 2006: 70). If there is a genuine drone war, then combat drones could be understood as a form of enhanced information warfare. Its usage remains embedded in highly developed information structures and worldwide communication networks – a direct continuation of developments in microelectronics and information technology. The moment somebody wants to destroy a drone, this person could also attack the computers and satellites – not physically but employing software in cyber operations. A drone war does not take place without an information or cyberwar.

Especially with respect to cyberwar, expected and traditional distinctions between combatant and non-combatant become blurry. This observation pertains to the usage of combat drones as well – not because drone technologies could replace human soldiers. That is not the point. The point is linked to the information networks behind global drone operations. Classical borderlines between criminality and war become blurry as cyberspaces generate battlespaces and spaces for crime at the same time that are at the moment not adequately regulated on a legal level. Actors in cyberspaces could be private persons, criminal hackers, government safety devices, members of military units (classical combatants) or economic companies – each one of them with specific motives and power. Clarifying the notion of these aspects in the domain of cyberwar will also touch other domains of enhanced information warfare, such as combat drones.

Another cultural impact is located on the level of public perception: minimum collateral damage, fewer civil victims, fewer human casualties in Western armies – those are some promises of the new war technologies like combat drones. But a consequence of promising military surgical precision from the perspectives of Western democracies is a change in the perception of casualties: every victim is one victim too many (Münkler 2006: 205f.). Politicians in Western democracies need to find strategies of legitimization for every single killed individual. Those strategies are shaped by rules of (mass) media presentation. News produces the perception of war for the public. War becomes an audiovisual experience. So people become emotionalized, and politicians search for strategies of public self-expression (Frech & Trummer 2005: 167–169). At the same time, the impact of national territory loses its main influence on warfare (Mey 2001: 87f.). Threats from long distance and dangers related to cruise missiles, weapons of mass destruction, terrorism, international crime, and strategic information related operations become more important (Mey 2001: 103–110). Combat drones belong to that development such as cyberwar. The developmental path of microelectronics and information technologies enables many social effects and technical applications. Drones are one effect of it, and they need to be analyzed in a broader context of media representation, actions in cyberspaces, or global security politics.

Especially from the point of view of Western democracies, current military technological developmental paths cannot be fundamentally departed from. That is true for air superiority, digital battles, robots, and drones, as well as for small but well-trained quick reaction forces in the context of war against terrorism (Irrgang 2014b: 159f.). If there are principal gaps, they remain economic because technical and economic development goes hand in hand; that is another characteristic of those developmental paths. On the one hand, high-tech warfare could become cheaper. For example, unmanned combat systems are expected to have up to 80% less overhead (Petermann & Grünwald 2011: 151). On the other hand, within drone warfare, the criterion of success is changing. The aspect of cultural embedding loses its impact as specific bodily skills on the battlefield become secondary. Emerging from that perspective are at least two cultural impacts that can be differentiated. The first one is directly related to "culture" in the sense of the "tradition of knowing". As information structures become globalized and the knowledge of using those structures, such as the knowledge of interpreting visual screen pictures and handling joysticks, traditional handcraft-oriented borderlines between different cultures of knowledge in the bodily handling of weapons become blurry. Bodily interface knowledge (between operator and joystick, screen, keyboard, etc.), of course, is not suspended, but the criterion of military success changes, as, with respect to combat drones, economic issues become much more dominant. If somebody is not able to pay for constructing, producing, and applying a high-tech drone with its complex information networks, then it is not so easy to compensate for that lack with a better technique, better competences, or more elaborated skills. But if you have better high-technology knowledge, more satellites, better engineers and scientists with enough money and resources to do their jobs efficiently in your institutes, then you are in a superior position. Even if someone has the best nerds, who are the most skilled in playing computer games, those competences would lie idle without economic power to buy or develop drone technologies and their networks. Again, the criterion of military success changes in that the economic and technological impact is not necessarily related simply to the mass production of drones. Success does not mean having the most drones. Success means better information technologies to ensure precise operations of the best drones operating in a worldwide information network. The characteristic of economic influences is changing from mass production to specialized case-oriented high technology (Irrgang 2008).

Symbolic for that development is the "revolution in military affairs". The introduction of so-called intelligent weapons, the enhanced precision and marksmanship while using long-distance weapons, and the acceleration of information exchange on the battlefield have brought about the military superiority of the United States. Military equipment becomes more complex and expensive to purchase. The number of actors who are able to participate in these high-tech wars decreases. In the end, the only nation with capabilities of global warfare activities remains the United States (Frech & Trummer 2005: 17–19; the same is true with respect to combat drones. At the same time, more and more worldwide conflicts can be observed, which are characterized by the usage of simple and cheap weapons such as machine guns, converted pickups, and the like (Frech & Trummer 2005: 18–20). Economic and technological differences between low-budget wars and high-tech warfare become bigger. Low-budget wars are less globalized and more strongly culturally embedded; "techniques" and skills have a stronger influence on military success. That is one reason why the criterion for the success of military operations differs between both domains. Combat drones are the symbols and flagships of high-tech warfare. The use of those technologies cannot be understood in terms of classical handcraft- or mass-produced-oriented military conflicts. It is enhanced information warfare and somehow enhanced electronic warfare as well – a military symbol for a general change from industrialized mass production societies to information and service societies. The drone becomes a techno-morph combat-service provider.

We don´t think that stopping the drone war completely is the most important ethical point here. Drones as technical systems are deeply interwoven in many other technological structures, which have become a part of our everyday lives. We cannot leave our technological, social, and economic developmental paths so easily. That does not mean fatalism, but there are ethical challenges. To put it simply: drones don´t care about national, social, or cultural borderlines; drones don´t care about values and ethics; drones are tools. That´s why we need transnational and transcultural regulations – not because of the drones but because of the things *we do* with drones. Establishing those regulations seems to be the most important ethical challenge of drone war. Combat drone responsibility is not a matter of single nations. But because of the intercultural and worldwide usage and technological knowledge background of the usage of combat drones, it calls for a global framework such as the UN or maybe some other new institution.

We should not stress the point that both knowledge and responsibility belong to humans and that drones do not know anything. We should also not stress the point too much that human knowledge is not only propositional knowledge but also tacit knowledge, embodied knowledge and skills. We should stress more important ethical issues and find ways to generate competences and transnational frameworks in regulating the usage of drones! Ethical reflections and judgment are a specific human form of knowledge itself. Ethics does not necessarily mean duties or abstract rules. It means competences to find and justify practical decisions for concrete practical situations (Irrgang 2007). Ethics may be abstract and idealistic in the way of illustrating long-term goals, such as "no more wars". But without the competences needed to go in that direction *step by step*, goals like this will never be reached. What we need are idealistic aims and pragmatic skills.

4. Responsibilities: Towards Moral Practice

On global markets, drones and their technical subsystems are mostly available without regulation. According to Thomas Petermann and Reinhard Grünwald, the law of nations does not forbid the usage of armed drones per se. But a variety of necessary parameters need to be adapted, such as international harmonized jurisdiction, technical standards, and insurance models. On a global level, according to international law, a distant goal is the regulation of armed drone operations as well as the trade of dual-use devices (which could be used to build drones without legal control) (Petermann & Grünwald 2011: 13–18). Also, following Mary Kaldor, new wars need to be analyzed in the context of globalization processes (Kaldor 2007: 18). Globalization processes are an ambivalent phenomenon. There is also a contrast between, on the one hand, societies oriented in the international political order and the ideal of cosmopolitanism, including global companies, NGOs, and global scientific communities, and on the other hand, more nationally and traditionally defined religious or social groups that feel like minorities and isolated from the benefits of globalization (Kaldor 2007: 130f.). Those tensions call for political solutions, not for drone war. If combat drones are high-tech weapons used by a few global players, we must insist that drones are no justification for a one-sided law of the jungle. Means and ends are intrinsically intertwined, as Mark Coeckelbergh emphasizes:

> However, my point here is not to defend a particular view on the justification of war or of particular political and military doctrines, but to show how means and ends are intrinsically linked and to argue that therefore we should not dismiss a discussion of ends when analyzing and evaluating military robotics.
> (Coeckelbergh 2011b: 3)

Ethical standards are necessary to prevent the misuse of drones. We need an ethos, a moral codex for drone war. It always sounds elaborate to argue for abstract ethical duties or to require an ethos for something. In the end, the problem is always simple: "Yes! But how to implement it in political reality? How to realize it concretely in everyday life?"

One proposal is formulated by Robert Sparrow, who declares, "Just say no to drones" (Sparrow 2012). His appeal is related to the robotics community and the engineers working on those kinds of technologies because most of the financial support for that research comes from military sources. Sparrow wants engineers to think about what they are doing and how they are affected by military funding. Responsibility for combat drones is an interdisciplinary challenge not only for philosophers, politicians, and military personnel but also for technical experts, designers and engineers. Following Sparrow, one opportunity is to find alternative ways of funding that are independent from military sources.

> Hopefully most, if not all, engineers can agree, along with the rest of the community, that we would all be better served if, in the future, robots were being researched, designed, and built to confront some of the urgent social and environmental challenges facing humanity today rather than to kill or to wield political power in foreign lands.
>
> (Sparrow 2012: 63)

"Just say no to drones" – that is human autonomy (form 1). The technical embedded autonomy of drones (form 3) finds its limits exactly at this point where people "just say no to drones". It is a gift of democracy that we can say no, maybe not to robotic technologies as such, but to combat drones. We want to emphasize another argument. Maybe saying yes or no to drones is not the right question. Maybe the right question is related to the frameworks of using drones. Let´s be a little idealistic, which is always good for setting aims. Technological-economic developmental paths, as introduced it in the last section, are also social and cultural paths. Technical development does not mean necessarily technocracy or predetermined destiny. Material and economic issues are interwoven with social expectations or values. We should not forget that the same enlightened rationality that enabled natural-scientific and technological developments during the last 400 years is also a social, ethical, and political form of enlightenment. Without enlightenment, no drone technologies! This is an ambivalent assessment. Technical rationality played a major role also in two destructive 20th-century world wars. So somebody could argue that the same rationality and enlightenment today causes horrible worldwide drone wars. But this is not the point. The point is that we can learn from history. That is human autonomy (form 1). We must not reproduce former mistakes. The initial spark of current technologies in 16th- and 17th-century sciences and technologies was also an initial ignition for philosophical ideas of peace. There wouldn´t be any drone technologies without the modern idea of peace. We should carefully internalize that connection.

One more aspect of human autonomy: learning from mistakes and the results of the past to enfold rational and humane potentials for the future. One idealistic potential and goal is Perpetual Peace (*Zum ewigen Frieden*), related to an Idea for a Universal History with a Cosmopolitan Purpose (*Idee zu einer allgemeinen Geschichte in weltbürgerlicher Absicht*), as Immanuel Kant revealed it at the end of the 18th century. That is our tradition, and interwoven into the developmental paths from early modern technologies to current high-tech drones are also developmental paths of transnational organizations starting with Kant´s ideas, leading to the League of Nations and currently the UN. Here we touch on frameworks of today´s drone war responsibilities. This is the idealistic side of the coin: there are goals of peace and cosmopolitanism that promote transnational organizations. But at the same time this ideal should not cause arrogance and the misuse of information or technological-economic superiority. Humans are finite and culturally embedded beings. That is our autonomy and strength, and calls not for Euro-centrism or United States–centrism but for sensitivity of cultural pluralities. A pragmatic counterpart is necessary to realize idealistic aims step by step. We should not try to become immortal or infinite beings. Also, drones should not be seen as substitutes for that dream of living forever. We should assume the challenge of finite steps to reach a bigger goal. One of these steps is interdisciplinary work.

We support the appeal of Sparrow to make engineers think about their concrete responsibilities in times of military-funded robotics research. Following the concept of a hermeneutical ethic (*hermeneutische Ethik*), ethical pragmatism is a form of competence (Irrgang 2007). Learning ethical competences should be integrated in engineering curricula. That does not mean dogmatic indoctrination of saying yes or no. Ethical pragmatism and hermeneutical ethics provide a terminological and conceptual toolbox. Argumentative and reflexive usage of ethical tools such as utilitarianism or deontological ethics can enable pragmatic abilities to evaluate concrete situations. With those philosophical tools, engineers should be able to find their own reasons and arguments. Sharing ethical competence is an interdisciplinary challenge and a necessary step in these times of the combat drone. Much work needs to be done in both domains: developing and enhancing transnational legal, political, and ethical frameworks and interdisciplinary work that starts even in the education of technical practitioners. Last but not least, we want to formulate at least three fundamental moral claims (which could be complemented by others)[13] of a combat drone ethos:

1 We are morally responsible to be aware of our autonomy, to learn from history, and to find rational reasons and justified arguments for future developments. We should self-confidently apply our human autonomy (form 1), and we should not search for our autonomy in drones or behave like drones (form 3).
2 We are morally responsible to work interdisciplinarily (including not only engineers, computer scientists, or designers, who develop such tools, but also philosophers and politicians).
3 We are morally responsible to built up transnational frameworks for the regulation of development, trading, using, and otherwise dealing with combat drones.

Last but not least:

4 We are morally responsible to follow idealistic ethical aims and to learn pragmatic ethical skills. The competence of saying a justified no to combat drones belongs both to human moral responsibility and autonomy.

Notes

1 Here we agree with Alex Leveringhaus, who also argues in Chapter 9 of this volume that drones "signal both continuity and change".
2 That has been discussed in an increasing number of publications on new wars and on warfare generally: see, for example, Etzersdorfer (2007); Kaldor (2007); Münkler (2002); Münkler (2006); Roither (2010); and others.
3 For terminology and the technical state of the art, see for example Petermann & Grünwald (2011: 6ff., 106*ff.) or* Sparrow (2007: 62ff). Different kinds of drones (already existing or not yet existing) are also summarized by Ezio Di Nucci and Filippa Santoni de Sio in the Introduction of this book.
4 In context of robotics research, the impact of forms of life in human–robot interactions has recently been investigated in Coeckelbergh 2014.
5 Arguments like this have often been applied in the context of robotics research, especially with respect to humanoid or service robots (Behnke 2008: 5; Coeckelbergh 2011a: 200; Decker 2010: 46ff.; Decker 2014: 101; Funk 2014a: 70; Irrgang 2005: 167; Ishihara 2014: 47; Kemp et al. 2008: 1308; Nakao 2014).
6 The mirror-metaphor has often been used in current robotics debates from different points of view, also in the context of mirror neurons, second-person perspective and human–robot interactions (Coeckelbergh 2014: 63, 66; Funk 2014a: 70ff., 79ff.' Ihde 2010: 42; Irrgang 2014a: 39; Zimmerli 2014: 18).
7 In terms of accessible transparency, in this volume Tjerk de Greef (Chapter 7) discusses the circumstance that humans are responsible, not drones.
8 The problem of dual use and misuse is discussed by Aimee van Wynsberghe and Michael Nagenborg in this volume (Chapter 8).
9 This observation is a leading paradigm also for philosophical investigations about drones (Christaller et al. 2001: 127; Decker 2014; Funk 2014a: 80f.; Irrgang 2005: 165, 212).
10 The tool character of technical artifacts and its cultural embedding in human practices have been prominently investigated in 20th-century philosophy of technology (Borgmann 1984; Ihde 1990; Irrgang 2001; Mitcham 1994).
11 A philosophical theory of technological-economical developmental paths has been developed in Irrgang (2002, 2008).
12 Distance perspectives are also discussed by Bernhard Koch in this volume (Chapter 4).
13 The duties of drone operators, as discussed by Jesse Kirkpatrick in Chapter 5 in this volume, may be a fruitful addition to our approach.

References

Behnke, S. 2008. Humanoid Robots – From Fiction to Reality? *Künstliche Intelligenz* 22(4): 5–9.

Borgmann, A. 1984. *Technology and the Character of Contemporary Life. A Philosophical Inquiry*. Chicago: University of Chicago Press.

Christaller, T., M. Decker, J.-M. Gilsbach, G. Hirzinger, K. Lauterbach, E. Schweighofer, G. Schweitzer, & D. Sturma. 2001. *Robotik. Perspektiven für menschliches Handeln in der zukünftigen Gesellschaft*. Berlin & Heidelberg: Springer.

Coeckelbergh, M. 2011a. Humans, Animals, and Robots: A Phenomenological Approach to Human-Robot Relations. *International Journal of Social Robotics* 3(2): 197–204.

———. 2011b. From Killer Machines to Doctrines and Swarms, or Why Ethics of Military Robotics Is Not (Necessarily) about Robots. *Philosophy & Technology* 24(3): 269–278.

———. 2013a. Too Close to Kill, Too Far to Talk. Interpretation and Narrative in Drone Fighting and Surveillance in Public Places, in R. Leenes & E. Kosta (eds). *Bridging Distances in Technology and Regulation*. Oisterwijk: Wolf Legal Publishers, 125–133.

———. 2013b. Drones, Information Technology, and Distance: Mapping the Moral Epistemology of Remote Fighting. *Ethics and Information Technology* 15(2): 87–98.

———. 2014. Robotic Appearances and Forms of Life. A Phenomenological-hermeneutical Approach to the Relation between Robotics and Culture, in M. Funk & B. Irrgang (eds). *Robotics in Germany and Japan. Philosophical and Technical Perspectives*. Frankfurt am Main: Peter Lang, 59–68.

David, P.A. 1985. Clio and the Economics of QWERTY. *American Economic Review* 75(2): 332–337.

Decker, M. 2010. Ein Abbild des Menschen: Humanoide Roboter, in M. Bölker, M. Gutmann, & W. Hesse (eds). *Information und Menschenbild*. Berlin/Heidelberg: Springer, 41–62.

———. 2014. Who Is Taking Over? Technology Assessment of Autonomous (Service) Robots, in M. Funk & B. Irrgang (eds). *Robotics in Germany and Japan. Philosophical and Technical Perspectives*. Frankfurt am Main: Peter Lang, 91–110.

Etzersdorfer, I. 2007. *Krieg. Eine Einführung in die Theorien bewaffneter Konflikte*. Wien: Böhlau Verlag.

Frech, S., & P. Trummer. 2005. *Neue Kriege. Akteure, Gewaltmärkte, Ökonomie*. Oldenburg: Wochenschau Verlag.

Funk, M. 2012. Leibbilden, Notenbilden, Onlinebilden – Wie wissen wir Musik zu teilen?, in W. Sützl, F. Stalder, R. Maier, & Th. Hug (eds). *Media, Knowledge and Education: Cultures and Ethics of Sharing. Medien – Wissen – Bildung: Kulturen und Ethiken des Teilens*. Innsbruck: Wochenschau Verlag, 209–226.

———. 2014a. Humanoid Robots and Human Knowing – Perspectivity and Hermeneutics in Terms of Material Culture, in M. Funk & B. Irrgang (eds). *Robotics in Germany and Japan. Philosophical and Technical Perspectives*. Frankfurt am Main: Peter Lang, 69–87.

———. 2014b. Robots and Responsibility. A Reply to Mark Coeckelbergh, in J. Seibt, R. Hakli, & M. Nørskov (eds). *Robo-Philosophy – Social Robotics and the Future of Social Relations. Conference Proceedings*. Aarhus: IOS Press, 271–276.

Funk, M., & M. Coeckelbergh. 2013. Is Gesture Knowledge? A Philosophical Approach to the Epistemology of Musical Gestures, in H. De Preester (ed). *Moving Imagination – Explorations of Gesture and Inner Movement in the Arts*. Amsterdam & Philadelphia: John Benjamins, 113–132.

Ihde, D. 1990. *Technology and the Lifeworld. From Garden to Earth*. Bloomington: University of Indiana Press.

———. 2010. Phenomenologists and Robots, in D. Ihde (ed). *Embodied Technics*. Copenhagen: Automatic Press Publishing, 37–53.

Irrgang, B. 2001. *Technische Kultur. Instrumentelles Verstehen und technisches Handeln*. Philosophie der Technik Bd. 1. Paderborn.

———. 2002. *Technische Praxis. Gestaltungsperspektiven technischer Entwicklung*. Philosophie der Technik Bd. 2. Paderborn.

———. 2005. *Posthumanes Menschsein? Künstliche Intelligenz, Cyberspace, Roboter, Cyborgs und Designer-Menschen – Anthropologie des künstlichen Menschen im 21. Jahrhundert*. Stuttgart: Franz Steiner.

———. 2007. *Hermeneutische Ethik. Pragmatisch-ethische Orientierung in technologischen Gesellschaften*. Darmstadt: WBG (Wissenschaftliche Buchgesellschaft).

———. 2008. *Philosophie der Technik*. Darmstadt: WBG (Wissenschaftliche Buchgesellschaft).

———. 2014a. Robotics as a future vision for hypermodern technologies, in M. Funk & B. Irrgang (eds). *Robotics in Germany and Japan. Philosophical and Technical Perspectives*. Frankfurt am Main: Peter Lang, 29–43.

———. 2014b. Drohnen und Kampfroboter. Neue Militätechnik für den gerechten Krieg im Globalisierungsstrudel? *Ethica* 22(2): 139–161.

Ishihara, Kohji. 2014. Roboethics and the Synthetic Approach – A Perspective on Roboethics from Japanese Robotics Research, in M. Funk & B. Irrgang (eds). *Robotics in Germany and Japan. Philosophical and Technical Perspectives*. Frankfurt am Main: Peter Lang, 45–58.

Kaldor, M. 2007. *Neue und alte Kriege. Organisierte Gewalt im Zeitalter der Globalisierung*. Frankfurt am Main: Suhrkamp.

Kemp, Ch. S., P. Fitzpatrick, H., H. Hirukawa, K. Yokoi, K. Harada, & Y. Matsumoto. 2008. Humanoids, in B. Sicilliano & P. Khatib (eds). *Springer Handbook of Robotics*. Berlin & Heidelberg: Springer, 1309–1333.

Mey, H. 2001. *Deutsche Sicherheitspolitik*. Frankfurt am Main: Report Verlag.

Mitcham, C. 1994. *Thinking Through Technology. The Path between Engineering and Philosophy*. Chicago: University of Chicago Press.

Münkler, H. 2002. *Die neuen Kriege*. Reinbek bei Hamburg¨Rowohlt Taschenbuch Verlag.

———. 2006. *Der Wandel des Krieges. Von der Symmetrie zur Asymmetrie*. Weilerswist: Velbrück Wissenschaft.

Nakao, M. 2014. Robots in Japanese Popular Culture, in M. Funk & B. Irrgang (eds). *Robotics in Germany and Japan. Philosophical and Technical Perspectives*. Frankfurt am Main: Peter Lang, 113–123.

Petermann, Th., & R. Grünwald. 2011. *Stand und Perspektiven der militärischen Nutzung unbemannter Systeme*. TAB-Arbeitsbericht Nr. 144. Berlin: TAB-Arbeitsbericht.

Roithner, Th. 2010. *Söldner, Schurken, Seepiraten. Von der Privatisierung der Sicherheit und dem Chaos der "neuen" Kriege*. Münster: LIT.

Sparrow, R. 2007. Killer Robots. *Journal of Applied Philosophy*, 24(1): 62–77.

———. 2012. 'Just Say No' to Drones. *IEEE Technology and Society Magazine* Spring: 57–62.

Wittgenstein, L. 2006. Philosophische Untersuchungen, in L. Wittgenstein (ed). *Werkausgabe Band 1*. Frankfurt am Main: Suhrkamp, 225–577.

Zimmerli, W. Ch. 2014. From Fiction or Science: A German-Japanese Era–project, in M. Funk & B. Irrgang (eds). *Robotics in Germany and Japan. Philosophical and Technical Perspectives*. Frankfurt am Main: Peter Lang, 11–25.

11 Autonomous Killer Drones

Nikil Mukerji

In this chapter, I address the question of whether autonomous combat drones should be allowed to make life-and-death decisions. To this end, I examine an argument proposed by Rob Sparrow, in which considerations of moral responsibility occupy centre stage. Though I reject his argument, I agree both with Sparrow's conclusion and with his basic contention that the idea of responsibility should play an important role in the investigation of the problem at hand. Therefore, I propose a different argument to show that we should not allow autonomous robots and more specifically autonomous drones to make life-and-death decisions. This argument also invokes the idea of responsibility. However, it focuses on a slightly different concept, viz. that of *attributive* or *agential* responsibility, which is closely connected with the idea of agency in a philosophically substantial sense. My argument assumes the intuitive principle that morally significant choices should only be made by subjects who are capable of responsibility in this sense, which can be reconciled both with a deontological and a consequentialist view of morality. Since killer robots and in particular autonomous killer drones seem to lack that capability, it follows that they should not be entrusted to make life-and-death decisions.

Robotic systems have increasingly become a part of our modern world. For quite some time now, they have played a great role in industrial production, where they have taken over repetitive processes that they can perform better, quicker, and more cheaply than human operators used to. Care-providing robots like FRIEND (*F*unctional *R*obot arm with user-fr*IEN*dly interface for *D*isabled people)[1] assist elderly and disabled persons in their daily lives. Surgical robots like da Vinci[2] support surgeons in performing their duties. They extend limited human capacities and are able to compensate for certain human weaknesses (e.g. trembling). All these uses of robotic technology open up important and contentious ethical questions that we as a society have to answer one way or the other. Industrial robots have, for example, arguably taken over people's jobs and have thereby endangered their livelihoods (e.g. Mukerji 2014a). Care-providing robots raise, for instance, the question of whether the people cared for will develop a sense of objectification and whether, ethically speaking, this is an acceptable price to pay, should this turn out to be true (e.g. Sharkey and

Sharkey 2012). Surgical systems raise issues of justice, for example, if they are available to only a select few (Palmerini et al. 2014: 90). All of these issues certainly represent pressing and important concerns. We should not treat them lightly. It seems, however, that these problems are dwarfed by the severity of the ethical issues associated with the use of robotic systems in military combat. The reason is, I think, obvious. Industrial robots may take people's jobs. Care robots may conceivably objectify the people they care for. And by deploying scarce surgical systems, we may quite possibly treat some people unfairly. But military robots and drones in particular can *kill* people, if we allow them to do that. In fact, some of them are designed with that very motive in mind,[3] and it is well known that the U.S. military, in particular, has used them to that end multiple times (e.g. in Pakistan and Yemen). Hence, it seems fair to argue that the subject matter of military robots and drones in warfare is not 'just another' complex of issues in the ethics of technology. When it comes to this particular issue, the stakes are very high and demand the careful attention of ethicists, scientists, politicians, and citizens.

In this chapter, I seek to contribute to the debate about military robotic technologies by discussing the moral credentials of what I call autonomous killer drones (AKDs). To this end, I proceed as follows. In the first section, I briefly comment on the state of the debate about AKDs before I examine, in the second section, an argument by Rob Sparrow that rejects AKDs based on considerations of responsibility. In the third section, I analyse Sparrow's argument and reject it, though I am inclined to agree with its conclusion, viz. that the use of AKDs is morally impermissible. For this reason, I propose, in the fourth section, the sketch of a new argument that also employs the idea of responsibility in order to establish Sparrow's conclusion. In the final section, I sum up and conclude with a few thoughts on the status of my argument for the overall debate about the moral justifiability of AKDs.

1. The Debate on Autonomous Killer Drones

In this chapter, I seek to examine whether the use of autonomous killer drones (AKDs), as I call them, is morally permissible. Since the introduction to this volume already provides a helpful discussion of drone technologies and likely developments, I do not have to go into too much detail about the technological aspects of these devices. So I will confine myself to reemphasizing that AKDs are specific types of unmanned aerial vehicles (UAVs), or drones, that carry deadly weaponry and are not only unmanned but also *autonomous* (as opposed to *automated* – a concept that is discussed in Chapter 9 of this volume by Alex Leveringhaus). Conventional UAVs need a human being 'in the loop'; that is, they need a person who controls their operations and steers them. More advanced versions are able to carry out specific tasks of medium complexity without a human intervention, requiring merely some form of monitoring by a human operator on the loop. AKDs are characterized by an even greater degree of autonomy. They need not be controlled or even monitored by human operators but can act in a rather autonomous fashion. They can also learn from past acts and events and from the data they regularly gather.[4]

A number of arguments for and against the use of AKDs in combat have been proposed and are discussed elsewhere in this volume.[5] Their proponents have pointed out, for example that AKDs can help our military save costs[6] and avoid putting human soldiers in harm's way (e.g. Swinson 1994).[7] They have also drawn our attention to the fact that, in virtue of being machines, AKDs are more precise (e.g. Arkin 2014; Sparrow 2009; Strawser 2010), most importantly because they are not prone to specific sources of error that are typical of humans. Humans are emotional and may overreact in certain situations, while AKDs will presumably keep a cool head (or processor, rather) when they react to, say, an unprovoked attack. In addition, proponents of AKDs may submit that these devices can operate without human intervention, which may be strategically advantageous. As Coeckelbergh (2013) points out, UAVs (and specifically AKDs) morally remove humans from the battlefield[8] and reduce, therefore, the threshold for military missions. It therefore becomes psychologically easier to engage in war, and threats to enemy forces become more credible and effective. On the other hand, critics of AKDs may press a number of objections that speak against their use in warfare. A rather general one may be based on a strictly pacifist view,[9] according to which killing is generally impermissible and specifically so when it is done through the use of AKDs. Of course, this is not the only point that critics can make. They might even agree that there are legitimate reasons to engage in warfare, while holding, based on consequential concerns, that we should nevertheless fight our wars in the conventional way. To this end, they might use Coeckelbergh's (2013) aforementioned observation that AKDs morally remove warfighters from the battlefield and point out that, therefore, they may plausibly cause more death and destruction.[10] A further, obvious argument that critics may employ focuses on the risks of AKDs as a new technology. They can make this plausible empirically by drawing attention to the numerous past malfunctions in drones. Back in 2003, the accident rate of U.S. military drones was reportedly 100 times that of manned aerial vehicles (Bone and Bolkcom 2003). And the accident rate of today's U.S. military drones is still much higher than that of manned aerial vehicles (Bloomberg 2014).

All these considerations are important when it comes to the overall evaluation of the moral credentials of AKDs. My aim in this paper is, however, much more modest. In what follows, I seek to investigate how considerations about responsibility may be used to make a case against AKDs. Therefore, even if my argument is successful, it will not show, in and of itself, that we should not use AKDs. It will merely show that there is *one* good reason not to use them. Furthermore, I should point out that much of what I will say is tentative. In other words, my argument does not stand on a rock solid foundation. I simply believe and will try to show that it appears quite plausible and worth taking seriously – at this stage anyway. With that in mind, I will now turn to a specific argument about AKDs and responsibility that has been advanced in the recent past.

2. Sparrow's Argument

Rob Sparrow was one of the first authors to address the moral issues involved in automated robotic warfare. His argument against autonomous robotic killing machines – or "killer robots", as he calls them – does not specifically target AKDs. It includes all kinds of killer robots. However, since AKDs are a particular kind of killer robot, Sparrow's argument does apply to them in particular.

Sparrow starts off by distinguishing between two possible meanings of the word "autonomous". Oftentimes, he explains, "the claim that a weapon is 'autonomous' means only that it is capable of acting independently of immediate human control", which, as he goes on to elaborate, typically "means that it is a 'fire and forget' system capable of determining its trajectory or pursuing its target to some limited extent" (Sparrow 2007: 65). The use of this type of weapon system may be objectionable. But Sparrow thinks that it does not raise any new ethical issues. Future autonomous weapon systems, however, may be of a different type and may in fact raise new concerns. Once deployed, they may be able to act autonomously "in some more robust sense", which is to say that they may be able to make and pursue their own decisions about, say, a target to be eliminated. The notion of a *decision* may seem metaphorical because Sparrow is talking about machines, which are not typically considered to be capable of agency. But it is clear that he intends it in a more or less literal sense when he says that the decisions of a weapon system are "responsive to the internal states – 'desires', 'beliefs' and 'values' – of the system itself" (Sparrow 2007: 65). Sparrow clarifies, furthermore, that the system "will have significant capacity to form and revise these beliefs" and that it will hence be able "to learn from experience" (Sparrow 2007: 65).

Then Sparrow asks what or who could be held responsible if an autonomous weapon system committed a war crime. To this end, he paints the following scenario in which an AKD:

> directed by a sophisticated artificial intelligence, deliberately bombs a column of enemy soldiers who have clearly indicated their desire to surrender. These soldiers have laid down their weapons and pose no immediate threat to friendly forces or non-combatants. Let us also stipulate that this bombing was not a mistake; there was no targeting error, no confusion in the machine's orders, etc. It was a decision taken by the [AKD] with full knowledge of the situation and the likely consequences. Indeed, let us include in the description of the case, that the [AKD, N.M.] had reasons for what it did; perhaps it killed them because it calculated that the military costs of watching over them and keeping them prisoner were too high, perhaps to strike fear into the hearts of onlooking combatants, perhaps to test its weapon systems, or because the robot was seeking to revenge the 'deaths' of robot comrades recently destroyed in battle. However, whatever the reasons, they were not the sort to morally justify the act. Had a human being committed the act, they would immediately be charged with a war crime.
>
> (Sparrow 2007: 66)

Given that the AKD in Sparrow's scenario commits what under normal circumstances would be called a war crime, the question naturally arises who, if anyone, can be held responsible for it. Sparrow explains why this question matters. "The question of the attribution of responsibility in this situation matters", he argues, "because . . . it is a fundamental condition of fighting a just war that someone may be held responsible for the deaths of enemies killed in the course of it" (Sparrow 2007: 67).

Now Sparrow proceeds by arguing that, in the preceding scenario, it seems as though nobody can plausibly be attributed responsibility for the deaths caused by the AKD. To show this, he considers three obvious alternatives: the person who developed and programmed it, the AKD itself, and the officer who ordered its use. Eventually, he rules out all of these possibilities.

Sparrow admits that it is "tempting to insist that the fault lies with the person(s) who designed and/or programmed the weapon", as some have suggested (e.g. Kuflik 1999). However, he concludes that it would not be fair to do this unless the programmers acted negligently in some way. He gives us two reasons why he thinks that. His first reason is that the programmers may have made it clear to the purchaser of the weapon that unwanted acts like these cannot be excluded. If, this disclaimer notwithstanding, the weapon is used anyway, the programmers cannot be held responsible for it. Rather, the responsibility is assumed by those who then decided to use the weapon anyway. His second reason is more fundamental and draws on the idea of autonomy. The possibility "that an autonomous system will make choices other than those predicted and encouraged by its programmers", argues Sparrow, "is inherent in the claim that it is autonomous." And he goes on to say that "[i]f it has sufficient autonomy that it learns from its experience and surroundings then it may make decisions which reflect these as much, or more than, its initial programming" (Sparrow 2007: 70).

The second solution is to hold the commanding officer who ordered the use of the AKD responsible.[11] Sparrow concedes that this may sound plausible because "[t]he risk that [the AKD, N.M.] may go awry is accepted when the decision is made to send it into action" (Sparrow 2007: 70). He rejects it nevertheless because there is the "risk that military personnel will be held responsible for the acts of machines whose decisions they did not control". This, Sparrow holds, would not be fair either because an agent cannot be held responsible for factors that are beyond her control.

Thirdly, there is the option to hold the machine itself responsible, which Sparrow rejects as well.[12] To this end, he asks "what would follow from holding it to be responsible" (Sparrow 2007: 71). In his view, to hold somebody responsible is, first and foremost, to affirm that they are "the appropriate locus of blame or praise and consequently for punishment or reward" (Sparrow 2007: 71). Then Sparrow goes on to explain that the application of the concept of holding someone responsible depends on the possibility of that agent being punished or rewarded. In the case of a machine, Sparrow does not see how this should be possible.

3. Why Sparrow's Argument Fails

Vincent Müller criticizes Sparrow's argument in his contribution to this volume (Chapter 3). In what follows, I will do so too. However, my critique is different from Müller's in two ways. Firstly, as I said in the beginning, my beef with Sparrow's argument lies not in the conclusion. Unlike Müller, who welcomes – tentatively anyway – the advent of AKDs, I tend to side with Sparrow in rejecting them. So my critique is not hostile to Sparrow's cause. Secondly, I focus on different issues than does Müller. While they offer a partners-in-guilt argument – *viz.* that we should accept the responsibility gap that Sparrow laments because we do so in other areas of the ethics of engineering – I will criticize Sparrow for relying on a view of the concept of responsibility that, I think, can reasonably be rejected. I suspect that my reasons for doing so will seem a bit idiosyncratic to my particular view of responsibility. Unfortunately, I cannot fully justify it here but will merely be able to explain the basic rationale behind it.[13]

The conception of responsibility that I advocate is in large part simply the result of taking the "control principle" (Mukerji and Lütge 2014; Nelkin 2013) seriously. This principle constrains the scope of objects for which an agent can reasonably be seen as responsible.[14] It states that anything outside the control of the agent cannot reasonably be attributed to her as the responsible party. Much could be said about what follows from this idea. For present purposes, however, I shall confine myself to a couple of remarks that appear important in view of my investigation.

First up, I should emphasize a distinction between two concepts of responsibility that Sparrow appears to run together. Responsibility may be understood, firstly, in the sense of attributability and, secondly, in the sense of substantive responsibility (Mukerji 2014b). When we investigate responsibility in the former sense, we deal, in the first instance, with a factual question, *viz.* whether a given object can be attributed to someone as the agent to whom it can be ascribed. We employ the notion of responsibility in the second sense, such as when we describe a given act as responsible or irresponsible. In doing this, we do, of course, express an evaluation that has implications for the appropriateness of certain "reactive attitudes" (Strawson 1962). Responsible acts should be praised, while irresponsible acts should be met with resentment.[15] It is important to note that these two senses of responsibility are to some extent independent. When we seek to identify an agent as responsible for a given act, no evaluation needs to be involved. When we ask for example, "Who pushed the button?" we are trying to figure out to whom the act of pushing the button can be attributed. When we say, on the other hand, that it was irresponsible to push the button because a sign clearly stated that this would set off a bomb, we are not attributing the pushing of the button to an agent. We are simply saying that whoever did it acted in a way that violated a moral prohibition. Now there is evidence that Sparrow conflates these two senses of responsibility when he invokes the principles of *jus in bello*. As he argues, these principles demand that when an AKD causes an atrocity, we should be able to identify someone who assumes responsibility. Here he clearly uses the idea of responsibility in the sense of attributability. Then, however, he discusses whether it would be fair or unfair to ascribe responsibility to particular agents, *viz.* the programmer and the commanding officer. This consideration, however, is neither here nor there. It is, in itself, neither fair nor unfair to attribute responsibility to an agent. It is simply a *factual* claim that may be true or false. The fact that Sparrow finds it relevant to ask the fairness question indicates that evaluative questions about substantive responsibility are mixed into his thinking at the wrong point.

Perhaps this seems a bit pernickety. It may be argued that, this conceptual confusion notwithstanding, Sparrow may be construed as having a valid substantive point. After all, it still seems true that when an AKD kills, the deaths it causes cannot be attributed to any agent as the responsible party. This, in turn, would make it impossible to blame and punish someone for the deaths caused by the robot, even though blame seems to be called for. At this point, I would like to back up a little and make a second basic remark about the concept of responsibility.

As Nida-Rümelin (2014a) argues, it is a conceptual implication, in conjunction with the control principle, that agents are responsible for *all* their acts. An act, after all, is a reason-guided form of behaviour. We cannot speak of an act unless the agent in question *decides* to do it. In other words, acts are deliberate.[16] Hence, all acts fall within the sphere of agential control, such that it follows, by the control principle, that an agent is responsible for them. Accordingly, it is always proper to ascribe attributive responsibility for an act to its respective agent. Whenever an agent acts, her act can be attributed to her as *her* act. However, on the control principle, the consequences of an agent's acts cannot reasonably be regarded as falling within the scope of objects for which an agent may be responsible.[17] Consequences, after all, are not entirely controlled by the agent, since they do not depend only on the act itself but also on the environment in which it is done. The latter, of course, is not entirely controlled by the agent. To be sure, agents do control the *probability* distribution over a set of potential consequences, which should be seen as part of the description of her respective options (Savage 1954/1972). And they may appropriately be seen as responsible for doing an act that is associated with a particular probability distribution, such that the *anticipated* consequences may influence the moral appraisal of the act. But the *actual* consequences that result from her acts are not objects of responsibility in the view I advocate.

At this point, I can come back to the issue at hand. It may still seem problematic, as I said, that the deaths of individuals caused by the involvement of an AKD cannot be attributed to any agent. If what I said in the previous paragraph is accepted, however, the question as to who may take responsibility for these deaths is a misguided question to begin with. These deaths should properly be regarded as the (unfortunate) *consequence* of the act (or acts) of an agent (or agents). As such, nobody should be seen as responsible for them. But this does not mean that when AKDs kill, we are faced with a moral vacuum. A number of agents can be regarded as responsible for creating a situation in which the AKD could do what it did. A chain of acts made this possible, and each part of the chain can, of course, be evaluated in moral terms. If, indeed, there are decisive moral reasons to reject the use of AKDs in warfare, then we can ask, for instance, whether the programmer acted irresponsibly by making the weapon available to the military. And we can ask whether those who ordered the AKD's use acted irresponsibly in doing so. Again, if the use of AKDs is morally wrong, then any programmer who assists in developing them and any officer who chooses to use them act wrongly and are blameworthy for it. Sparrow's worries thus seem unfounded. When AKDs are ordered into action and kill, we can hold someone responsible for that.[18]

4. A New Argument from Responsibility

At this stage, it is, of course, an open question whether those who contribute to the use of AKDs in warfare deserve to be blamed. This depends on whether the use of these weapons is, in fact, morally wrong. In this section, I would like to introduce, explain, and defend, to some extent anyway, a new argument to this effect that, like Sparrow's, relies on considerations of responsibility but is more congenial with my view of responsibility. In concise form, it can be stated thus:

(P1): If an entity is not capable of acting in the philosophically relevant sense, it cannot be a subject of responsibility.
(P2): AKDs are not capable of acting in the philosophically relevant sense.
(P3): If an entity cannot be the subject of responsibility, then it should not be entrusted to make morally significant choices and act on them.
(P4): Life-and-death decisions are morally significant choices.
(C): AKDs should not be entrusted to make life-and-death decisions and act on them.

Let me first explain the premises of the argument. P1 says that, if an entity is not capable of acting in the philosophically relevant sense, it cannot be a subject of responsibility. This claim hooks up with the account of responsibility that I have already introduced. When we say that an agent does a given act and if we mean that in the philosophically interesting sense on which my previous remarks focused, we ascribe to her *eo ipso* a responsibility for that act. Conversely, when we attribute to an agent the responsibility for something, that something has to be an aspect of her reason-guided behaviour. Agency and responsibility, in other words, are "two sides of the same coin" (Nida-Rümelin 2014a: 24).

P2 says that AKDs are not capable of acting in the philosophically relevant sense. To be sure, the claim is not that they do not behave in certain ways. When they engage and attack a target, they undoubtedly exhibit some form of behaviour. The claim is merely that this behaviour is not accompanied with specific mental processes that characterize humans, for instance, and turn them into agents in the philosophically interesting sense.

P3 states that if an entity cannot be the subject of responsibility, then it should not be entrusted to make morally significant choices and act on them.[19] To explain, a morally significant choice occurs when the behaviour of a given entity affects certain normative factors and is hence relevant from the perspective of moral theory. For instance, all reasonable moral theories agree that the suffering of sentient beings is bad and should be avoided. A situation where the suffering of sentient beings is at stake would thus constitute a morally significant choice. The claim that P3 makes is that the choosing done within such a choice should not be left to a mindless entity that lacks the mental processes that are necessary for agency in the relevant sense.

Finally, P4 claims simply that a situation where some entity can make a life-and-death decision is a morally significant choice in the sense just explained.

As far as I can see, the logical side of the argument for C from P1–P4 is rather unsuspicious (e.g. the argument is deductively valid, contains no equivocations, and so on). The tenability of the argument should thus come down to the credentials of the premises. So what can be said to justify them? As already stated, I will not be able to provide a watertight reasoning for the immorality of AKDs. Rather, I am aiming only to provide the sketch of an argument.

P1 and P4, I take it, wear their credentials on their sleeves. P4 appears self-evident and needs no further justification. And P1 appears to be conceptually true. It definitely is on the account of responsibility that I have previously advocated and which ties this idea to the idea of agency. But even those who would reject some of the conceptual commitments that I would endorse should, it seems, accept P1. After all, this premise merely claims that when no agent is present and things merely happen, as it were, then there is no responsibility. To be sure, events that are tragic from a moral point of view may occur in such situations. When an earthquake devastates a village and causes a lot of suffering among its inhabitants, for instance, this is morally bad because all reasonable moral theories would regard the suffering of innocent people as bad. But it is equally clear that, with no agent present, such a tragedy does not fall within the responsibility of anyone.[20] This leaves me with P2 and P3, which I cannot discuss in nearly enough detail here. The best I can do, I think, is to show that there is some reason to believe them – for now anyway.

My reason for thinking that P2 is true – at this point in time – is simply this: it has not been shown that computer-operated devices can act in ways that at least resemble, let alone replicate, the ways in which a human agent acts. For now, computers simply seem to lack the ability to perform the mental processes that are required for real agency. This, to me, is simply a matter of empirical observation. I suspect, however, that some would find it unsatisfactory if I stopped my discussion of P2 here. So let me expand a bit.

I might start by observing that there are two camps of theorists that are divided on P2. Some find it very commonsensical – which, of course, is not to say that it is *true* or that it will remain so for all eternity. However, they think that it seems at least *prima facie* inconceivable, at this point in time, that a machine – like an AKD – would be capable of mental processes, let alone the precise kinds of mental processes that characterize agency in the proper sense. This, at any rate, is roughly in line with what some of the foremost thinkers of the modern age have concluded (e.g. Leibniz and Descartes).[21] And to the extent that there is currently no evidence to the contrary, one may find it justified to cautiously endorse P2 – at least for now. The second camp is more optimistic about machine agency, precisely because they see evidence to the contrary. To support their optimism, they may point to the so called Turing Test that Allen Turing famously suggested as a test for machine intelligence.[22] In the Turing Test, a machine is tasked to converse with a human being it is digitally connected with. The point is for the machine to persuade its interlocutor that it is an actual person. Turing himself thought that by the year 2000 it would be possible to programme machines "to make them play the Imitation Game so well that an average interrogator will not have more than 70 per cent chance of making the right identification after five minutes of questioning" (Turing 1950, 442). In 2014 – 14 years later than Turing had prophesized – the first computer programme, the chatbot Eugene Goostman, was publicly announced to have cleared that mark. Posing as a little Ukrainian boy, the programme convinced 33% of the judges that it was an actual person (BBC 2014). Theorists in the second camp may thus hold themselves justified in believing that machines can think and that this undermines P2.

It would be hasty to conclude, however, that this proves the second camp right. There are a number of reasons why one might want to take a sceptical attitude. For one thing, the claim that Eugene Goostman passed the Turing Test got a lot of heat (e.g. Mann 2014) because the programme posed as a 13-year-old non-native speaker and simply dodged many of the questions it was asked. So it is questionable whether it should really be regarded as having passed the Turing Test. A more basic point, however, is that it is not clear whether the Turing Test even is an adequate test for the existence of a mental life in the required sense. In this connection, Searle's (1980) famous Chinese Room thought experiment comes to mind, which seeks to discredit the idea that a computer programme – even if it did pass the Turing Test – would develop a real understanding of the meaning of a conversation. Rather, it would merely mimic, mindlessly, the formal procedure of the thought process behind it. To show this, Searle dissects the way a computer supposedly "thinks", viz. by manipulating symbols according to certain rules. He develops a scenario in which he, a human operator, acts in ways that exactly parallel a computer's behaviour. That is, he imagines himself sitting in a room, where he is given Chinese symbols on a piece of paper from outside the room. In reply, he squiggles down certain other Chinese symbols, following a set of rules provided to him, and passes them back. If the rules are sufficiently sophisticated, Searle argues, a person who speaks Chinese would, simply by looking at the exchange of symbols, not be able to tell that Searle is not a competent Chinese speaker. But, sure enough, he is not. He does not even remotely understand the meaning of the symbols he encounters or writes down. If they are interpreted by an actual Chinese speaker as a thoughtful exchange, then that is not Searle's doing. Since a computer, conversing with a human, would do exactly the same as Searle in his Chinese room scenario, it too is not able to develop a genuine understanding of the conversation that is taking place. Accordingly, one might argue, in line with what the first camp believes, that a computer cannot think after all – even if this may be our impression. To the extent that the ability to think is a necessary condition for agency in the philosophically relevant sense, which seems very plausible, any computer-operated machine (including an AKD) will necessarily fail to achieve it.

But this line of reasoning is problematic for a number of reasons, as critics may point out. First of all, Searle's argument is, in itself, quite controversial.[23] Furthermore, one may point out, as do Wendell Wallach and Colin Allen, that Searle's notion of genuine understanding "marks a distinction without a behavioral difference" (Wallach and Allen 2009, 58). Based on that, one may argue that a computer-operated device, like an AKD, can act in the philosophically relevant sense of the word "act", viz. to the extent that its behaviour would, for all practical intents and purposes, be indistinguishable from the behaviour of an actual agent. It would be goal-directed, and the device would possess the capacity to give (what we can interpret as) reasons for it. The fact that the machine may, for all we know, be a zombie-like entity with regard to its mental life may then be taken to be irrelevant.

Proponents of the first camp, however, may have another ace up their sleeves. They may argue that a computer-operated device would, in principle, never be able even to imitate a proper agent. As Deborah Johnson points out:

> [h]istorically human beings have been understood to be different from all other living entities because they are *free* and have the capacity to act from their freedom. Human beings can reason about and then *choose* how they behave.
>
> (Johnson 2006, 198; emphasis added, N.M.)

This capacity to choose freely, the ability to deliberate over a choice and to decide according to one's own well considered preferences and priorities, may set humans apart not only from other living beings. It may also set them apart from computers and machines. Arguably, the latter may never achieve that ability. If, after all, their decision procedure follows an algorithm, the choice they make would never be free but would always be predetermined (Nida-Rümelin 2014b). Of course, it is possible to include an element of randomness into an algorithm. But this would hardly do justice to what we call "freedom". It should be remarked, however, that a number of well renowned thinkers (e.g. Dennett 1997/2014; Kurzweil 2005; Sandberg and Bostrom 2008) have held vastly different views, founded on different ideas about the reconcilability of (algorithmic) determinism and agential freedom (see, in particular, Dennett 1984, 2003).[24] They think that it is possible to implement essentially the same processes that are characteristic of human agents in a machine (which is the underlying premise behind the enterprises of "whole brain emulation" and "mind uploading").

I, for one, cannot see any principled reasons why machines should not, at some stage, develop the capacity to choose freely and to act in a philosophically substantive sense. We know that it is possible to implement the processes that are required for this in a carbon-based substratum. So it seems to me to be an open *empirical* question as to whether it is possible to do the same in a silicon-based substratum. Accordingly, we should investigate the matter empirically and should be weary of any armchair arguments that seek to convince us of the contrary. Armchair philosophers have, after all, often attempted to prove the possibility or impossibility of a thing, only to be utterly ridiculed afterwards. Think for example of Hobbes's proof for the squaring of the circle or of Hegel's alleged belief in the a priori impossibility of an eighth planet in our solar system. Nevertheless – and this brings me back to the point that I started with – it is hard to see any evidence that would undermine P2 *at this point in time*. And to that extent, I think it is provisionally justified.

Finally, let me address P3. It can be made plausible as follows. As already stated in the previous section, attributive responsibility and agency go hand in hand on my view of these concepts. It follows that if an entity is incapable of one, it is incapable of the other. That is, if the entity in question is incapable of attributive responsibility, it will be incapable of the mental processes that are required for agency in general and specific types of agency in particular. Hence, it will also be incapable of the kinds of reasoned decisions that are characteristic of *moral* agency. Now it seems *prima facie* reasonable to suspect that such an entity will not be sufficiently responsive to the morally relevant nuances of their 'choice' situations. Their behaviour may thus be expected to be at variance with important moral principles. To put such an entity in charge of a morally relevant decision and have it act on it therefore seems wrong.[25]

Sparrow (2007) provides a useful analogy that may make the case for P3 more intuitive and compelling. He compares the use of autonomous lethal weapons, including AKDs, with the use of child soldiers in warfare. This comparison is, I think, instructive. There are, of course, many reasons why we should object to sending children into combat. But one aspect that is surely important is that children are not fully capable of agency and, *eo ipso*, not fully capable of responsibility either. They undoubtedly are able to think. They are able to deliberate over a decision. And they are able to choose. But they do not have the capacity of fully responsible agents, *viz.* to arrive at a truly well reasoned decision. This, amongst other things, would make it wrong to have them make morally significant decisions – let alone decisions about life and death – and have them act on them. The reason for objecting to the use of AKDs that I am proposing here is the same. AKDs may be able to decide and act in a basic sense. But if they lack the requisite mental processes that are characteristic of *real* agency in a philosophically substantive sense, they are not able to reason through their decisions. When they kill, there is no real agent who calls the shots. Things, as it were, "just happen". And that is a very scary notion.

This, then, is the justification of P3 in a nutshell. Much more could be said, of course, in the way of expounding the philosophical foundations that underlie it.[26] But this would go far beyond what I am trying to do here. As I have said, my goal in this piece is simply to clarify my argument and to make it, at least to some extent, plausible. I think in what I said so far I have largely been able to live up to this rather modest goal. It occurs to me, however, that I should make one more point that may help to make my view clearer and, at the same time, to defend it further. To this end, I shall briefly comment on the difference between consequentialist and deontological moral views and how the ideas I have presented relate to this distinction.

Since moral philosophers often categorize their views as being either deontological or consequentialist in nature, it seems reasonable to ask whether the philosophical worry that is expressed by P3 is motivated by deontological or consequentialist concerns.[27] In fact, what I said up until now may be interpreted either way. The reason I gave to make it plausible that an entity that is not capable of attributive responsibility should not be entrusted to make morally significant choices and act on them was that this entity would likely not be sufficiently receptive to the morally relevant nuances of a choice situation and its behaviour would, hence, likely violate important moral principles. This is deliberately vague and inconclusive between consequentialism and deontology. On a consequentialist rendering, my claim may be said to rest on the suspicion that putting such entities in charge of morally relevant choices will likely lead to bad consequences and should thus be avoided. One might prove this by showing that the entity in question would not be able to distinguish, add, and weigh the pleasures and pains of sentient beings, the degree to which their preferences are satisfied or frustrated and so on – depending on how the respective consequentialist perspective is fleshed out. On a deontological interpretation, it may be seen as a matter of principle not to put a mindless entity in charge of a morally relevant situation. This could then be proven, say, by arguing that doing so would express a lack of respect for the people who are affected by the outcome of the choice or that it would likely violate individual rights – again, depending on how the respective view is spelled out. Though I am, of course, not indifferent with respect to the controversy between consequentialists and deontologists, I believe that, in the present context, it would not be helpful to tailor my reasoning either to the deontological view or the consequentialist one. Rather, I consider it a virtue of my argument that moral theorists on both sides of the deontology/consequentialism divide can reconcile their views with it.

Conclusion

In this chapter, I examined and rejected Sparrow's argument against the use of AKDs. Sparrow (2007) claims that when AKDs kill, nobody can be held responsible for the deaths that result, which he takes to be a violation of the principles of *jus in bello*. On account of responsibility that I introduced, this worry is, however, unfounded. I argued that agents are always responsible for what they do but never responsible for the consequences of their acts. Hence, when AKDs kill, we can hold individuals responsible for ordering them into action, though we cannot hold them responsible for the deaths that are caused. But we can hold them responsible for the role they played in providing and deploying the weapon. If using AKDs is, indeed, morally wrong, people who through their individual acts contributed to their use acted wrongly and are responsible for that. All hinges, however, on whether or not it is, in fact, morally wrong to use AKDs. To provide evidence for this idea, I proposed a new argument, which employed the principle that morally significant decisions should be made only by proper agents, who are in possession of the mental faculties to reason them through. At this point in time, there is no indication that AKDs do possess these faculties, such that we may conclude, by my argument, that their use is morally wrong. As training and doctrine capabilities manager of the Future Combat Systems programme of the U.S. Army, Colonel Lee Fetterman, puts it, "The function that robots cannot perform for us – that is, the function we should not allow them to perform for us – is the *decide* function" (quoted by Davis 2007, emphasis added, N.M.).

To conclude, I shall briefly comment on the status of my argument. Have I shown that the use of AKDs is morally wrong, *period*? I surely hope that I have been able to provide some evidence for that view. But it would obviously be quite foolish for me to claim that I have established it conclusively. There are, broadly speaking, two reasons for this.

Firstly, I should note what I tried to make clear as I went through premises P2 and P3 of my argument, viz. that the foundations on which the argument rests are not entirely certain. In particular, the claim that AKDs are unable to act in a sense in which this would be required to ensure that they can, like human beings, be subjects of responsibility is not set in stone. At this stage, it would be narrow-minded to claim anything more than that they seem, as yet, incapable of agency in a sufficiently exacting sense. But a couple of centuries, decades, or even just years down the road, all of this may change.

Secondly, as I made clear in the first section, many aspects need to be taken into consideration when the moral credentials of AKDs are examined. These are for example considerations about the costs of war, the harm to our soldiers, the plights of civilians, strategic concerns, worries about the risks of AKDs, and so on. To be sure, I did not go into any depth about these issues. And for all I know, the points I mentioned may be completely flawed. Nevertheless, it has to be conceded that there are numerous arguments on both side of the debate that may conceivably support the pro or the con side. And these arguments may be stronger than the one I have proposed, such that the balance of reasons will actually support the use of AKDs in war. I doubt that that is the case. At this point in time, I surely tend to believe that their use is morally unjustifiable. But I have to concede that all I have been able to do here is to give *one* argument that supports my present view. The debate is still young. And it is fair to say that the jury is still out on AKDs.

Notes

1 Details on FRIEND can be found online at the University of Bremen website http://www.iat.uni-bremen.de/sixcms/detail.php?id=1078 (accessed 29 January 2016).
2 More information about Da Vinci can be found at http://www.intuitivesurgical.com/products (accessed 29 January 2016).
3 The MQ-1B (Predator), MQ-9 (Reaper) and MQ-1C (Gray Eagle), e.g. have the ability to carry Hellfire missiles.
4 Note that my argument is relevant only on the supposition that AKDs will be developed at some stage in the future. This possibility is doubted, for example by Asa Kasher in his contribution to this volume (Chapter 6). In my view, however, it is quite likely. However, whether this will turn out to be true depends on a number of factors that I will not try to examine here (e.g. technological possibilities, political and legal factors, and societal forces). For detailed analyses of the possibilities of drone technologies, see e.g. Singer (2009). For an analysis of relevant political and legal factors, see e.g. Benjamin (2013, chap. 6), De Shaw (2014, chap. 3), and Kaag and Kreps (2014, chap. 4). And for an examination of societal forces, see Benjamin (2013, chaps. 8 and 9) and Kaag and Kreps (2014, chap. 3).
5 A helpful overview is also given by Lin and colleagues (2009).
6 Drone technologies do, of course, cost a fortune in R&D and procurement (see e.g. GAO 2013). But humans who fly military missions in manned aircrafts need to be trained and paid for as well. And unlike a decommissioned drone, a human pilot is entitled to pension payments or disability benefits after service.

7 See also Autonomous UAV Missions Systems (AUMS), Spawar at http://www.public.navy.mil/spawar/Pacific/Robotics/Pages/AUMS.aspx (accessed 29 January 2016).

8 A good case for this can be made based on the neuropsychological study of trolley cases (e.g. Greene et al. 2009), which suggests that people tend to accept the sacrifice of a person's life for the sake of a higher purpose, if the person who is sacrificed is at some physical distance from the agent. Whether that intuition can be justified is a contested issue in general moral philosophy. Singer (1972) for example has argued for the view that the physical distance of a needy stranger does not affect our duty to help her (see also Unger 1996). Kamm (2007) has famously opposed this view.

9 Modern variants of this moral stance can be found in, e.g., Holmes (1999) and Fox (2013).

10 This point has been made by various authors in the academic debate, such as by Sparrow (2009), but also in the public forum, such as by the Nobel Peace Laureate Jody Williams (Russia Today 2014).

11 Sparrow attributes this idea to Lazarski (2002).

12 This idea is advocated, e.g., by Beck (2014).

13 See Mukerji (2014b) for further remarks and Nida-Rümelin (2011, 2014a) for a more comprehensive defence.

14 Note that Sparrow himself seems to adhere to some version of the control principle. After all, his reason for not holding the programmer/officer responsible for atrocities committed by autonomous robots has to do with the fact that their behaviour is beyond the control of the programmer/officer. Proponents of the idea of "moral luck" (Nagel 1979 and Williams 1982) would, of course, deny the close connection between agential control and responsibility.

15 Whether a given act should be called "responsible" is, of course, a substantive moral question that depends on what the best moral theory is. If utilitarianism is the best theory, then an act should be seen as responsible if it maximizes overall happiness. Similarly, if a particular deontological theory is the most preferable one, then acts that transgress certain deontological restrictions (e.g. "Don't lie", "Don't kill", etc.) will have to be called "irresponsible" and so on. For present purposes, however, I can put these basic issues in moral theory aside.

16 Of course, it is nevertheless possible for an agent to do an act unintentionally. She may, after all, do something unintentionally *by* doing something else intentionally. However, it is not possible for an agent to perform an unintentional act if she does not perform an intentional act by which that unintentional act is done (see Anscombe 1957).

17 Note that the claim makes sense only if we can distinguish between an act and its consequences. As has been pointed out by Oldenquist (1966) and subsequently by a number of further authors, this may be problematic. It is, of course, to some degree arbitrary whether we describe what happens when an agent acts as a part of her act or as a part of her act's consequences. What I do may, for example "be described variously as making marks on a piece of paper, signing a cheque, paying a bribe, or ensuring the survival of my business" (Sumner 1987: 166). However, in the given context, there seems to be a criterion. When somebody acts, we can specify that anything in that situation that is under the reason-guided control of the agent belongs to the act and anything that does not (or not fully) is a consequence, such that it becomes possible to interpret the idea in a meaningful way.

18 There may, however, be a separate issue here. The individual responsibilities of the responsible parties and the blame we may justifiably ascribe to them may, in a sense, not "add up" to the appropriate amount. This well known complication occurs in "many-hands problems" (Thompson 1980), where many agents interact. A well known example is traffic. Many people die in traffic each year as a result of the unfortunate interaction of many individuals. Oftentimes we are unable to identify anyone in particular who should justifiably take the full blame. As Hellström (2013) remarks, "Instead, an underlying agreement is that the advantages with our

traffic systems are larger than the disadvantages". And he goes on to explain that "[t]he same way of reasoning could in principle be applied to unpredictable, but appreciated, autonomous war robots" (Hellström 2013: 105–106). Whether this idea presents a satisfactory solution to the many-hands problem in the case of AKDs is something I shall not assess here.

19 The senses in which the words "make", "choices", and "act" are used here are, of course, different from the senses in which they are used when we describe a human agency. In this case, their meaning is metaphorical.

20 This does not rule out, of course, that someone may be responsible for certain aspects of such a situation that call for moral assessment. For example, when an earthquake destroys a village, we may ask who was responsible for putting it there in the first place?

21 Descartes, in particular, provided a thought experiment to that effect in his *Discourse on the Method* (1637), which evidently is a forerunner of the famous Turing Test (Oppy and Dowe 2011).

22 In fact, Turing (1950) suggested his test as, so to speak, "the next best thing" because he believed that the question of whether machines can think was "too meaningless to deserve discussion" (Turing 1950, 442).

23 See e.g. Cole (2014), for an overview of the debate about the Chinese Room thought experiment.

24 In Dennett's case, agential freedom and determinism are reconciled through a compatibilist view of free will. In his view, humans are themselves nothing other than "moist robots" (Dennett 2014, 49).

25 I should point out that I am not assuming here that we need to make sure that the entity in question will always behave in the right way. Rather, what seems important is that we can recognize that it is receptive to certain moral reasons and distinctions that we can accept as relevant (see Wallach and Allen 2009, 70).

26 See, however, Nida-Rümelin (2011, 2014a) with whom I largely agree.

27 When presenting this argument, I have been asked on which side of the deontological/consequentialist divide it falls. I am especially indebted to Filippo Santoni de Sio and Ezio Di Nucci for pressing me on that point.

References

Anscombe, G.E.M. 1957. *Intention.* Oxford: Oxford University Press.

Arkin, R.C. 2010. The Case for Ethical Autonomy in Unmanned Systems. *Journal of Military Ethics* 9(4): 332–341.

Arkin, R.C. 2014. Lethal Autonomous Systems and the Plight of the Non-combatant. *Ethics and Armed Forces* 2014(1): 3–10.

BBC. 2014. *Computer AI Passes Turing Test in 'World First'*, 9 June. Available online at http://www.bbc.co.uk/news/technology-27762088 (accessed 29 January 2016).

Beck, S. 2014. Dealing with the Diffusion of Legal Responsibility: The Case of Robotics, in F. Battaglia, N. Mukerji, and J. Nida-Rümelin (eds). *Rethinking Responsibility in Science and Technology*. Pisa: Pisa University Press, 167–182.

Benjamin, M. 2013. *Drone Warfare: Killing by Remote Control*. New York: Verso Books.

Bloomberg 2014. *Drones Have Highest Accident Rates*. Available online at http://www.bloomberg.com/chart/inHAM4os9yZA (accessed 29 January 2016).

Bone, E., and C. Bolkcom. 2003. *Unmanned Aerial Vehicles: Background and Issues for Congress (Report for Congress – Order Code RL31872)*. Available online at http://fas.org/irp/crs/RL31872.pdf (accessed 29 January 2016).

Coeckelbergh, M. 2013. Drones, Information Technology, and Distance: Mapping the Moral Epistemology of Remote Fighting. *Ethics and Information Technology* 15(2): 87–98.

Cole, D. 2014. The Chinese Room Argument, in E.N. Zalta (ed). *The Stanford Encyclopedia of Philosophy* (Summer). Available online at http://plato.stanford.edu/archives/sum2014/entries/chinese-room (accessed 29 January 2016).

Davis, D. 2007. Who Decides: Man or Machine? *Armed Forces Journal* 1 November. Available online at http://www.armedforcesjournal.com/who-decides-man-or-machine (accessed 29 January 2016).

Dennett, D. 1984. *Elbow Room: The Varieties of Free Will Worth Wanting*. Cambridge, MA: MIT Press.

———. 2003. *Freedom Evolves*. London: Penguin Books.

———, D. 1997/2014. When HAL Kills, Who's to Blame?, in F. Battaglia, N. Mukerji and J. Nida-Rümelin (eds). *Rethinking Responsibility in Science and Technology*. Pisa: Pisa University Press.

———. 2014. *Intuition Pumps and Other Tools for Thinking*. New York: W. W. Norton & Company.

De Shaw, J. 2014. *Analyzing the Drone Debates: Targeted Killings, Remote Warfare, and Military Technology*. New York: Palgrave Macmillan.

Fox, M.A. 2013. *Understanding Peace: A Comprehensive Introduction*. London: Routledge.

GAO 2013. *Defense Acquisitions – Assessments of Selected Weapon Programs* (March 2013), U.S. Government Accountability Office. Available online at http://www.gao.gov/assets/660/653379.pdf (accessed 29 January 2016).

Greene, J.D., F.A. Cushman, L.E. Stewart, K. Lowenberg, L.E. Nystrom, and J.D. Cohen. 2009. Pushing Moral Buttons: The Interact between Personal Force and Intention in Moral Judgment. *Cognition* 111(3): 364–371.

Hellström, T. 2013. On the Moral Responsibility of Military Robots. *Ethics and Information Technology* 15(2): 99–107.

Holmes, R. 1999. Pacifism for Nonpacifists. *Journal of Social Philosophy* 30(3): 387–400.

Johnson, D.G. 2006. Computer Systems: Moral Entities but Not Moral Agents. *Ethics and Information Technology* 8: 195–204.

Kamm, F. 2007. *Intricate Ethics: Rights, Responsibilities, and Permissible Harm* (Oxford Ethics Series). Oxford: Oxford University Press.

Kuflik, A. 1999. Computers in Control: Rational Transfer of Authority or Irresponsible Abdication of Autonomy. *Ethics and Information Technology* 1(3): 173–184.

Kurzweil, R. 2005. *The Singularity Is Near*. New York: Penguin Books.

Lazarski, A. 2002. Legal Implications of the Uninhabited Combat Aerial Vehicle. *Aerospace Power Journal* 16(2): 74–83.

Mann, A. 2014. That Computer Actually Got an F on the Turing Test. *Wired,* 6 September. Available online at http://www.wired.com/2014/06/turing-test-not-so-fast (accessed 29 January 2016).

McMahan, J. 2004. The Ethics of Killing in War. *Ethics* 114(4): 693–733.

Mukerji, N. 2014a. Why Moral Philosophers Should Watch Sci-fi Movies, in F. Battaglia and N. Weidenfeld (eds). *Roboethics in Film*. Pisa: Pisa University Press, 79–92.

Mukerji, N. 2014b. Technological Progress and Responsibility, in F. Battaglia, N. Mukerji, and J. Nida-Rümelin (eds). *Rethinking Responsibility in Science and Technology*. Pisa: Pisa University Press, 25–36.

Mukerji, N., and C. Lütge. 2014. Responsibility, Order Ethics, and Group Agency. *Archiv für Rechts – Und Sozialphilosophie* 100(2): 176–186.

Nagel, T. 1979. Moral Luck, in T. Nagel (ed). *Mortal Questions*. Cambridge: Cambridge University Press, 24–38.

Nelkin, D. K. 2013. Moral Luck, *The Stanford Encyclopedia of Philosophy* (Winter 2013 Edition), Edward N. Zalta (ed.), http://plato.stanford.edu/archives/win2013/entries/moral-luck.

Nida-Rümelin, J. 2011. *Verantwortung*. Stuttgart: Reclam.

———. 2014a. On the Concept of Responsibility, in F. Battaglia, N. Mukerji, and J. Nida-Rümelin (eds). *Rethinking Responsibility in Science and Technology*. Pisa: Pisa University Press, 13–24.

———. 2014b. Agency, Technology, and Responsibility. *Politica e Società* 2: 185–200.

Oldenquist, A. 1966. Rules and Consequences. *Mind* 75(298): 180–192.

Oppy, G., and D. Dowe. 2011. The Turing Test, in E. N. Zalta (ed). *The Stanford Encyclopedia of Philosophy* (Spring), Available online at http://plato.stanford.edu/archives/spr2011/entries/turing-test (accessed 29 January 2016).

Palmerini, E., F. Azzarri, F. Battaglia, A. Bertolini, A. Carnevale, J. Carpaneto, F. Cavallo, A. Di Carlo, M. Cempini, M. Controzzi, B.-J. Koops, . . . K. Warwick. 2014. *D6.2 – Guidelines on Regulating Robotics (Deliverable of the RoboLaw Research Project)*. Available online at http://robolaw.eu/deliverables.htm (accessed 29 January 2016).

Russia Today. 2014. *Autonomous 'Killer Robots' Could Replace Drones Soon' – Nobel Peace Prize Winner Jody Williams* (televised interview 13 January). Available online at http://youtu.be/iOrRX3ChdGg (accessed 29 January 2016)

Sandberg, A., and N. Bostrom. 2008. *Whole Brain Simulation – A Roadmap*. Technical Report #2008-3, Future of Humanity Institute, Oxford University. Available online at www.fhi.ox.ac.uk/reports/2008-3.pdf (accessed 29 January 2016).

Savage, L. 1954/1972. *The Foundations of Statistics* (2nd ed). New York: Dover Publications.

Searle, J. 1980. Minds, Brains and Programs. *Behavioral and Brain Sciences* 3: 417–457.

Sharkey, A., and N. Sharkey. 2012. Granny and the Robots: Ethical Issues in Robot Care for the Elderly. *Ethics and Information Technology* 14(1): 27–40.

Singer, P. 1972. Famine, Affluence, and Morality. *Philosophy and Public Affairs* 1(3): 229–243.

Singer, P.W. 2009. Wired for War: The Robotics Revolution and Conflict in the 21st Century. New York: Penguin.

Sparrow, R. 2007. Killer robots. *Journal of Applied Philosophy* 24(1): 62–77.

Sparrow, R. 2009. Predators or Plowshares? Arms Control of Robotic Weapons. *IEEE Technology and Society Magazine* 8(1): 25–29.77.

Strawser, B.J. 2010. Moral Predators: The Duty to Employ Uninhabited Aerial Vehicles. *Journal of Military Ethics* 9(4): 342–368.

Strawson, P. F. 1962. Freedom and Resentment. *Proceedings of the British Academy* 48: 1–25.

Sumner, L. W. 1987. *The Moral Foundation of Rights*. Oxford: Clarendon Press.

Thompson, D.F. 1980. Moral Responsibility of Public Officials: The Problem of Many Hands. *The American Political Science Review* 74: 905–916.

Turing, A. M. 1950. Computing Machinery and Intelligence. *Mind* 59: 433–460.

Unger, P. 1996. *Living High and Letting Die*. Oxford: Oxford University Press.

Wallach, W., and C. Allen. 2009. *Moral Machines*. Oxford: Oxford University Press.

Williams, B.A.O. 1982. Moral Luck, in B.A.O.Williams (ed). *Moral Luck*. Cambridge: Cambridge University Press, 20–39.

Index

For Product Safety Concerns and Information please contact our EU
representative GPSR@taylorandfrancis.com
Taylor & Francis Verlag GmbH, Kaufingerstraße 24, 80331 München, Germany

www.ingramcontent.com/pod-product-compliance
Ingram Content Group UK Ltd.
Pitfield, Milton Keynes, MK11 3LW, UK
UKHW022130190926
475702UK00013B/1728
9781138390669